Information and Instructions

This shop manual contains several sections each covering a specific group of wheel type tractors. The Tab Index on the preceding page can be used to locate the section pertaining to each group of tractors. Each section contains the necessary specifications and the brief but terse procedural data needed by a mechanic when repairing a tractor on which he has had no previous actual experience.

Within each section, the material is arranged in a systematic order beginning with an index which is followed immediately by a Table of Condensed Service Specifications. These specifications include dimensions, fits, clearances and timing instructions. Next in order of arrangement is the procedures paragraphs.

In the procedures paragraphs, the order of presentation starts with the front axle system and steering and proceeding toward the rear axle. The last paragraphs are devoted to the power take-off and power lift systems. Interspersed where needed are additional tabular specifications pertaining to wear limits, torquing, etc.

HOW TO USE THE INDEX

Suppose you want to know the procedure for R&R (remove and reinstall) of the engine camshaft. Your first step is to look in the index under the main heading of ENGINE until you find the entry "Camshaft." Now read to the right where under the column covering the tractor you are repairing, you will find a number which indicates the beginning paragraph pertaining to the camshaft. To locate this wanted paragraph in the manual, turn the pages until the running index appearing on the top outside corner of each page contains the number you are seeking. In this paragraph you will find the information concerning the removal of the camshaft.

More information available at haynes.com
Phone: 805-498-6703

Haynes Group Limited
Haynes North America, Inc.

ISBN 10: 0-87288-129-6
ISBN-13: 978-0-87288-129-7

MF-27, 4S1, 14-96

SHOP MANUAL

MASSEY-FERGUSON
MODELS

MF135 - MF150 - MF165

Tractor serial number stamped on instrument panel name plate.
Engine serial number stamped on side of engine.

INDEX (By Starting Paragraph)

C O N D E N S E D   S E R V I C E   D A T A

GENERAL	MF135 Special	MF135 Deluxe - MF150			MF165		
		Diesel	Non-Diesel	Non-Diesel	Diesel	Non-Diesel	Non-Diesel
Engine Make	Cont'l	Perkins	Cont'l	Perkins	Perkins	Cont'l	Perkins
Engine Model	Z-134	AD3.152	Z-145	AG3.152	AD4.203	G-176	AG4.212
Number of Cylinders	4	3	4	3	4	4	4
Bore—Inches	3.312	3.6	3.375	3.6	3.6	3.58	3.875
Stroke—Inches	3.875	5.0	4.062	5.0	5.0	4.38	4.5
Displacement—Cu. In.	134	152	145	152	203.5	176	212.3
Compression Ratio	6.6:1	17.4:1	7.1:1	7.5:1	18.5:1	7.1:1	7.0:1
Main Bearings, No. of	3	4	3	4	5	3	5
Cylinder Sleeves	Wet	Dry	Wet	Dry	Dry	Wet	Dry
Forward Speeds	6	6*	6*	6*	6*	6*	6*
Reverse Speeds	2	2*	2*	2*	2*	2*	2*

* 12 forward, 4 reverse; if equipped with "Multipower" Transmission.

TUNE-UP

	MF135 Special	MF135 Deluxe - MF150			MF165		
		Diesel	Non-Diesel (Continental)	Non-Diesel (Perkins)	Diesel	Non-Diesel (Continental)	Non-Diesel (Perkins)
Firing Order	1-3-4-2	1-2-3	1-3-4-2	1-2-3	1-3-4-2	1-3-4-2	1-3-4-2
Valve Tappet Gap (Hot)							
Intake	0.013	0.010	0.013	0.012	0.010	0.016	0.012
Exhaust	0.013	0.010	0.013	0.015	0.010	0.018	0.015
Compression Pressure @ Cranking Speed.................	140-145	....	145-160	....	....	145-160	....
Intake Valve Face Angle	30°	44°	30°	45°	44°	30°	45°
Intake Valve Seat Angle	30°	45°	30°	45°	45°	30°	45°
Exhaust Valve Face Angle	44°	44°	44°	45°	44°	44°	45°
Exhaust Valve Seat Angle	45°	45°	45°	45°	45°	45°	45°
Timing (Ign. or Inj.)							
Static	6° BTC	20° BTC	6° BTC	12° BTC	24° BTC	6° BTC	12° BTC
High Idle	30° BTC	....	26° BTC	26° BTC	....	26° BTC	36° BTC
Timing Location	Flywheel						C.P.
Battery							
Volts	12	12	12	12	12	12	12
Capacity Amp/Hr.	40	95	40	60	96	41	60
Ground Polarity	Neg.	Neg.	Neg.	Neg.	Neg.	Neg.	Neg.
Distributor Contact Gap	0.022	....	0.022	0.021	....	0.022	0.022
Spark Plug Size	14mm	....	14mm	14mm	....	14mm	14mm
Electrode Gap	0.025	....	0.025	0.025	....	0.025	0.025
Injectors							
Opening Pressure	....	2500 psi	....	....	2575 psi	....	....
Spray Hole Dia	....	0.0098	....	....	0.0098	....	....
Governed Speeds							
Engine							
Low Idle	475 rpm	700 rpm	475 rpm	750 rpm	750 rpm	475 rpm	750 rpm
High Idle	2225 rpm	2160 rpm	2225 rpm	2250 rpm	2160 rpm	2225 rpm	2250 rpm
Loaded	2000 rpm	2000 rpm	2000 rpm	2000 rpm	2000 rpm	2000 rpm	2000 rpm
Power Take-Off							
High Idle	707 rpm	686 rpm	707 rpm	708 rpm	686 rpm	707 rpm	708 rpm
Loaded	635 rpm	635 rpm	635 rpm	629 rpm	635 rpm	635 rpm	629 rpm
Horsepower @ PTO Shaft*..............	34.5	37.8	35.4	....	52.4	46.9	....
*According To Nebraska Tests							
Hydraulic System							
Maximum Pressure	See Paragraph 240						
Rated Delivery	4.5 or 4.85 gpm*						

*Depending on pump installed.

SIZES - CAPACITIES - CLEARANCES
(Clearances in Thousandths)

	MF135 Special	Diesel	Non-Diesel (Continental)	Non-Diesel (Perkins)	Diesel	Non-Diesel (Continental)	Non-Diesel (Perkins)
Crankshaft Journal Diameter	2.249	2.749	2.249	2.749	2.749	2.376	2.999
Crankpin Diameter	1.937	2.249	1.937	2.249	2.249	2.562	2.499
Camshaft Journal Diameter:							
Front	1.808	1.869	1.808	1.869	1.869	1.808	1.997
Center	1.746	1.859	1.746	1.859	1.859	1.746	1.987
Rear	1.683	1.839	1.683	1.839	1.839	1.683	1.967
Crankshaft Bearing Clearance:							
Main Bearings	1.5-2.5	3-5	1.5-2.5	3-5	3-5	1.5-2.5	2.5-4.5
Crankpin	0.3-2.3	2.5-4	0.3-2.3	2.5-4	2.5-4	0.7-2.7	1.5-3
Crankshaft End Play	4-8	2-15	4-8	2-15	2-14	4-8	2-15
Piston to Cylinder Clearance	1.5-2.5	5-7	1.5-2.5	....	5-7	1.5-2.5	....
Camshaft Bearing Clearance...........	2.5-4.5	4-8	2.5-4.5	4-8	4-8	2.5-4.5	2.5-5
Camshaft End Play....................	3-7	Sp. Loaded	3-7	Sp. Loaded	3-6	3-7	4-16
Cooling System—Quarts................	10	10.5	10	10.5	10.5	10.5	10.5
Crankcase—Quarts	6	6.5	6	6.5	8.5	6	8.5
Transmission, Differential and Hyd. Lift—Gal.	8	8	8	8	8	8	8
Power Steering—Qts.	⅔	1	⅔	1	1¾	1⅓	1⅓
Steering Gear Housing—Quarts	1	1	1	1	1	1	1
Planetary Final Drive	....	1¾*	....	....	1¾	1¾	1

*MF150 Diesel Row Crop only

FRONT SYSTEM

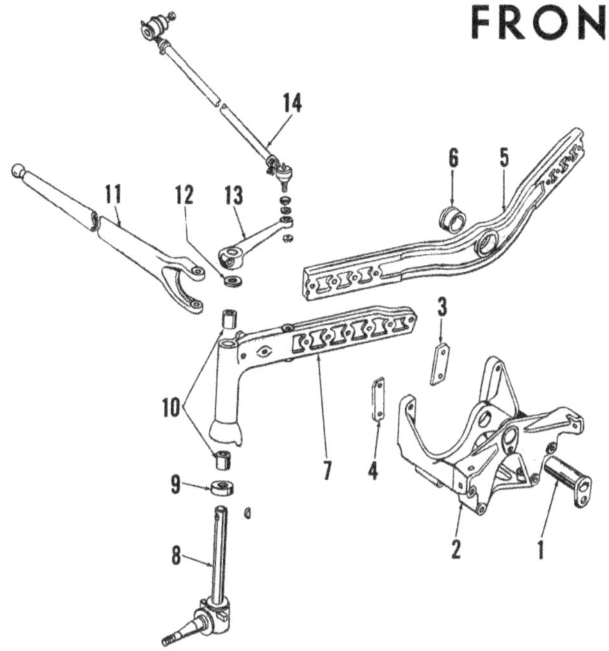

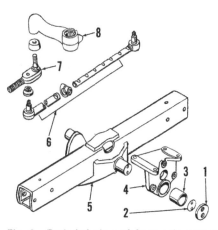

Fig. 1—Exploded view of front axle and associated parts used on Model MF-135.

1. Pivot pin
2. Front support
3. Spacer
4. Spacer
5. Axle center member
6. Bushing
7. Axle extension
8. Spindle
9. Bearing
10. Bushing
11. Radius rod
12. Dust seal
13. Steering arm
14. Drag link

Fig. 3—Axle pivot pin rear bushing is located in support housing as shown.

PIVOT BUSHING

Fig. 2—Exploded view of front axle center member and associated parts used on Model MF150.

1. Thrust plate
2. Shim
3. Bushing
4. Pivot bracket
5. Axle member
6. Tie rod
7. Tie rod end
8. Steering arm

MF135 tractors are available only in standard clearance, axle type. MF150 and MF165 units are available in standard clearance, axle type or high clearance (Row Crop) types with single or dual wheel tricycle or adjustable axle. Refer to the appropriate following paragraphs for service procedures.

AXLE ASSEMBLY

Model MF135

1. Refer to Fig. 1 for an exploded view of axle and support assembly. Axle extensions (7) may be removed by disconnecting radius rod (11) and drag link (14); then unbolting axle extension from axle center (5). To renew the axle center member (5), first support the tractor and remove lower pan from grille. Unbolt axle extensions (7) from center member; remove pivot pin (1) and withdraw center member (5) from front support.

A different front support (2) is used on gasoline and diesel models; however, diesel front support may be used on gasoline tractor by installing spacers (3 & 4).

All bushings are final sized and should not require reaming if carefully installed.

Model MF150

2. Refer to Fig. 2 for an exploded view of axle center member, steering tie rods and associated parts. Rear pivot bushing is located in front support casting as shown in Fig. 3; axle extensions and steering spindles are similar to those shown in Fig. 4.

To remove the axle main (center) member (5—Fig. 2), support tractor under engine, remove lower pan from grille and disconnect tie rods from spindle arms. Remove both axle extensions with wheels and spindles attached. Remove thrust plate (1) and save shims (2) for reinstallation. Remove the cap screws retaining pivot bracket (4) to front support casting and remove the pivot bracket. Pull center member (5) forward out of rear bushing.

Renew center member and/or bushings if clearance is excessive. Replacement bushings are pre-sized and will not require reaming if carefully installed. Make certain, however, that lubrication holes in bushings are aligned with fitting holes in castings. When installing rear bushing (Fig. 3), remove grease fitting (F) and stake bushing in place by inserting a slim punch through grease fitting hole and indenting bushing into depression in upper wall of bushing bore.

When installing axle center member, reverse the removal procedure and vary the number of shims (2—Fig. 2) to obtain an end play of 0.002-0.008 when checked between rear face of pivot bracket (4) and front face of center member (5). Shims are available in thicknesses of 0.002, 0.005 and 0.010.

Model MF165

3. Refer to Fig. 4 for exploded view of front axle and associated parts. Pivot pin (12) is retained in front support casting (15) by the cone point set screw (14).

Axle center member (9) or the complete axle assembly can be dropped downward out of front support after removing pivot pin (12). Bushing (8 & 10) and/or pivot pin (12) may be renewed at this time. When reinstalling, make sure the 3/32-inch thick thrust washer (11) is installed at rear of axle member (9), and vary the number and thickness of shims (7) to reduce end play to a minimum without binding. Shims (7) are available

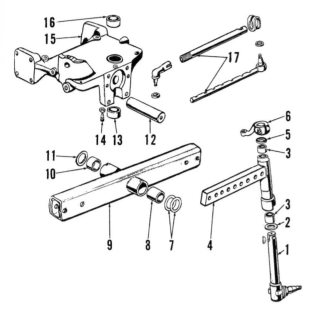

Fig. 4—Exploded view of
front support and axle assembly of the type used
on Model MF165.

1. Spindle
2. Thrust washer
3. Bushing
4. Axle extension
5. Dust seal
6. Steering arm
7. Shims
8. Bushing
9. Axle center member
10. Bushing
11. Thrust washer
12. Pivot pin
13. Bushing
14. Lock screw
15. Front support
16. Bushing
17. Tie rod

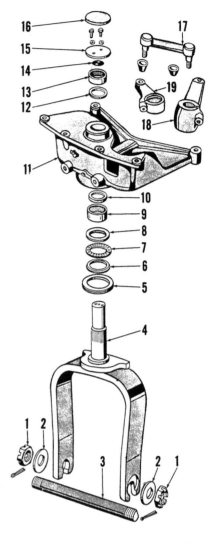

Fig. 5—Exploded view of single wheel fork,
support casting and associated parts used
on models so equipped.

1. Nut
2. Eccentric washer
3. Axle shaft
4. Wheel fork
5. Oil seal
6. Lower race
7. Bearing
8. Upper race
9. Needle bearing
10. Oil seal
11. Front support
12. Oil seal
13. Needle bearing
14. Shim
15. Thrust washer
16. Dust cap
17. Link
18. Steering arm
19. Steering arm

in thicknesses of 0.029, 0.035 and 0.040; shims should be installed at front of axle center member as shown.

SPINDLE BUSHINGS
All Models

4. Each axle extension contains two renewable bushings (10—Fig. 1 or 3—Fig. 4) which require final sizing after installation to provide the recommended 0.0035-0.005 clearance for the spindle. Nominal spindle diameter is 1½-inches for Model MF165 and 1¼-inches for other models.

TOE-IN, TIE-RODS AND/OR DRAG LINKS
All Models

5. Automotive type tie rod and drag link ends are used. Recommended toe-in is 0 - ¼ inch. On Model MF135, adjust both drag links an equal amount to obtain the recommended toe-in. On other models, remove the clamp bolt or loosen set screw on adjustable tie rod, and turn the adjusting sleeve in or out as required.

SINGLE WHEEL & FORK
All Models So Equipped

6. The fork-mounted single front wheel is carried in taper roller bearings which should be adjusted to provide a very slight rotational drag by removing the cotter pins and tightening or loosening the castellated spindle nuts (1—Fig. 5).

To remove the wheel fork, support tractor under engine and remove the grille lower panel. Remove the grille door and remove the sheet metal dust cap (16). Remove the retaining cap screws, thrust plate (15) and shim

pack (14). Working through the opening provided in support casting (11), loosen the clamp bolt retaining steering arm (19) to wheel fork. Raise front of tractor and at the same time, withdraw wheel fork (4) from below. CAUTION: Wheel fork must not be allowed to become cocked during removal, or seals and/or needle bearings may be damaged.

Examine needle bearings (9 & 13), seals (5, 10 & 12) and needle thrust bearing (6, 7 & 8) and renew any questionable parts. When installing caged needle bearings, be sure to align oil hole in bearing cage with the oil feed holes in support casting (11).

Install wheel fork by reversing the removal procedure. Vary the thickness of shim pack (14) to provide wheel fork with an end play of 0.002-0.008. Shims are available in thicknesses of 0.002, 0.005 and 0.010.

NOTE: The wheel fork and steering arm have a blind spline to facilitate correct assembly.

DUAL WHEEL SPINDLE & PEDESTAL
All Models So Equipped

7. Each of the dual wheels is mounted on taper roller bearings which should be adjusted to provide a very slight rotational drag.

To remove the lower spindle and wheels assembly, support tractor under engine and remove grille door and lower grille panel. Remove the cap screw (C—Fig. 6), tab washer and flat washer. Working through the opening provided in support casting (11), loosen the clamp bolt securing steering arm (1) to spindle (2), and

raise front of tractor while withdrawing spindle and wheels assembly from bottom of pedestal (7).

Lower pedestal (7) can be removed from support casting at this time if necessary for renewal of pedestal or bearing cups (6 & 8). Examine bearing cups, cones and seals, and renew any damaged or questionable parts.

Install by reversing the removal procedure, and tighten cap screw (C) to remove all spindle end play without causing bearing drag.

NOTE: Steering arm (1) and spindle (2) have a blind spline to facilitate correct installation.

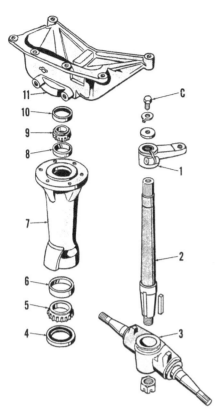

Fig. 6—Exploded view of dual wheel pedestal and associated parts.

C. Cap screw
1. Steering arm
2. Spindle
3. Wheel axle
4. Oil seal
5. Bearing cone
6. Bearing cup
7. Lower pedestal
8. Bearing cup
9. Bearing cone
10. Oil seal
11. Front support

Fig. 7—Left side view of steering gear housing used on Model MF 135, showing adjusting screw (A) and cap screw (L) which also serves as oil level indicator.

lubricant is maintained at level of left rear, side cover attaching cap screw (L—Fig. 7).

ADJUSTMENT

This section covers only steering gear backlash adjustment; and steering shaft bearing adjustment except on Model MF135 with power steering. For steering valve adjustment on power steering models, refer to ADJUSTMENT paragraphs of POWER STEERING SYSTEM Section.

STEERING GEAR

This section covers the mechanical steering gear and associated parts used on all models. On tractors with power assist steering, refer also to POWER STEERING SYSTEM section for information on power steering pumps, valves and cylinders. Model MF165 is equipped with power steering only; other models are optionally equipped with manual or power steering which use identical or similar gear units.

LUBRICATION

All Models

8. The recommended lubricant for steering gear on all models is SAE 90, mineral gear lubricant. Housing capacity is approximately one quart. Oil level and filler plug is located on left side of steering gear housing on MF150 and MF165 tractors. On Model MF135, filler plug is located on top, front of steering gear housing and

Model MF135

9. **BACKLASH ADJUSTMENT.** Teeth of sector gears (B & C—Fig. 8) are slightly tapered for backlash adjustment. Proceed as follows:

Loosen the locknuts and back out the adjusting screw (A) on each side of gear housing two full turns. Disconnect both steering drag links at rear end.

On power steering models, loosen and back out the Allen head cap screws retaining the power rack guide (42—Fig. 11) at least two turns.

On all models, working from right side of gear housing, tighten the right adjusting screw until all backlash is removed from LEFT sector arm, but gear does not bind when turned past center (straight-ahead) position. Adjust right sector gear in the same manner, working from LEFT side of housing. With both gears properly adjusted and drag links disconnected, rolling torque of steering wheel shaft measured at steering wheel rim should be approximately as follows:

For manual steering models—5 lbs.

For power steering models—2-2½ lbs.

On power steering models, retighten the power rack guide (42) after gears are adjusted; then, recheck rolling torque at steering wheel. If steering gear now binds or rolling torque is increased by more than ½ pound, re-

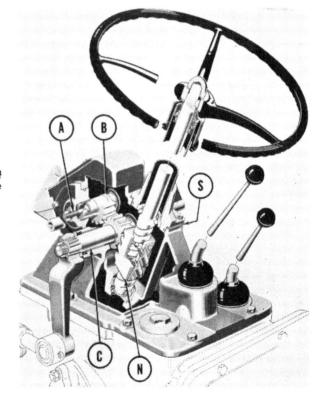

Fig. 8—Cutaway view of steering gear of the type used on Model MF135.

A. Adjusting screw
B. Sector gear
C. Sector gear
N. Ball nut
S. Adjusting shims

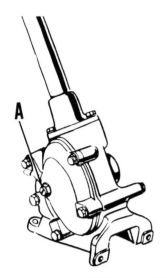

Fig. 9—On Models MF150 and MF165, turn adjusting screw (A) on housing cover, to adjust steering gear backlash.

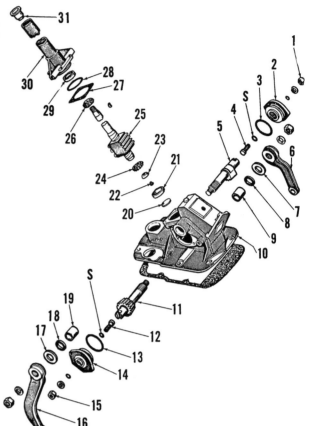

Fig. 10 — Exploded view of manual steering gear used on Model MF135.

1. Locknut
2. Side cover
3. "O" ring
4. Adjusting screw
5. Sector gear
6. Steering arm
7. Dust seal
8. Oil seal
9. Bushing
10. Housing
11. Sector gear
12. Adjusting screw
13. "O" ring
14. Side cover
15. Locknut
16. Steering arm
17. Dust seal
18. Oil seal
19. Bushing
20. Expansion plug
21. Bearing cup
22. Eyelet
23. Retainer
24. Bearing cone
25. Shaft & ball nut
26. Bearing cone
27. Shim
28. "O" ring
29. Bearing cup
30. Shaft housing
31. Bushing

move rack guide (42) and record the torque; then add shims (41) of sufficient thickness to increase torque 0-¼ lb. when rack is tightened. Shims are available in thicknesses of 0.003, 0.005 and 0.010.

Left sector gear MUST be adjusted first on all models, with right gear loosened. If wormshaft bearing end play exists on manual steering models, bearings should be adjusted as outlined in paragraph 10, before attempting to adjust backlash.

NOTE: On power steering models, do not confuse steering shaft end movement with gear backlash. A small amount of shaft movement must exist in order to actuate the steering valve.

10. **WORMSHAFT BEARING ADJUSTMENT.** The steering wormshaft ball bearings used on manual steering models should be adjusted to a slight preload by adding or removing shims (S—Fig. 8) located between steering shaft housing and gear housing. Steering shaft thrust load on power steering models is carried on thrust bearings at steering control valve and a slight amount of end movement is required to actuate the valve.

If only the bearings are to be adjusted without steering gear overhaul, disconnect both drag links at rear end and loosen both sector gear adjusting screws (A) two full turns as outlined in paragraphs 9. Remove the steering wheel and the four cap screws securing steering shaft housing to gear

housing; then add or remove shims (S) as required to remove all end and side play without binding. Preload is correct when a turning effort of ½-1½ lbs. (measured at steering wheel rim) is required to turn steering shaft with drag links disconnected and backlash adjustment loosened. Shims are available in thicknesses of 0.002, 0.005, 0.010 and 0.030 and are furnished only in a shim kit assortment.

Readjust backlash after bearing adjustment as outlined in paragraph 9. If a suitable adjustment cannot be obtained, overhaul the steering gear as outlined in paragraph 14.

Models MF150 - MF165

11. **BACKLASH ADJUSTMENT.** Steering gear backlash is adjusted by means of the adjusting screw (A—Fig. 9) located on right side of steering gear housing. Turning the adjusting screw clockwise reduces the backlash. Adjustment is correct when a barely perceptible drag or resistance is felt when steering wheel is turned through the mid-position, and no backlash exists. Before attempting to adjust the backlash, first check the steering wheel shaft (camshaft) bearings as outlined in paragraph 12, and adjust the preload if required.

12. **CAMSHAFT BEARINGS.** To check the camshaft bearings, first loosen the backlash adjusting screw (A—Fig. 9) at least two full turns. With adjustment screw loosened, check camshaft bearings for end play by pulling up and pushing down on steering wheel. Camshaft should turn freely with no perceptible looseness of bearings.

If end play exists, unbolt the steering column from gear housing and raise the column; then split and remove a sufficient quantity of shims (10—Fig. 14) to remove all end play. Shims are available in thicknesses of 0.002, 0.003 and 0.010. Adjust steering gear backlash as outlined in paragraph 11 after camshaft bearings have been adjusted.

OVERHAUL

Model MF135

The steering gear housing is combined with the transmission top cover and contains the gear shift levers and neutral safety switch. Refer to Fig. 10 for an exploded view of gear unit used on manual steering models and to Fig. 11 for unit used on power steering tractors. Most service on steering gear can be performed without removal of housing from tractor, however removal is required for service on transmission components. Refer to paragraph 13 for removal and installation of housing on all models; and to the appropriate paragraphs 14 through 15 for service, which may be performed as outlined, with or without housing removal.

13. **REMOVE AND REINSTALL.** To remove the steering gear housing for other work or to prepare tractor for steering gear overhaul, proceed as follows:

Remove the hood, battery, engine air cleaner and steering wheel. Disconnect the steering drag links at rear end. On diesel models, shut off fuel and disconnect fuel lines from filters on left side of battery support. On all models, block up underneath fuel tank and remove fuel tank rear support; or remove the tank. Remove the oil pressure feed line. Disconnect front and rear throttle control rods from cross shaft.

Disconnect fuel gage sender wire, the quick disconnect plug at starter switch, and the leads at the neutral safety switch. On models equipped with "Multi-power" transmission, disconnect the linkage and remove shift lever. On all models, remove the screws securing instrument panel to battery support and lay support forward on tractor, being careful not to damage the heat indicator capillary tube. Unbolt and remove the battery support.

Steering gear housing and transmission top cover assembly can be removed at this time by removing the securing cap screws; however, removal is not necessary for overhaul of the gear unit. Assemble and/or install by reversing the removal procedure. Bleed fuel system on diesel models as outlined in paragraph 136, and adjust "Multi-power" control linkage on models so equipped as outlined in paragraph 185.

14. **OVERHAUL MANUAL GEAR.** Refer to Fig. 10. To disassemble the steering gear, first remove steering arms (6 & 16) and the cap screws retaining side covers (2 & 14). Check splined ends of pinion shafts (5 & 11) and remove any paint, burrs or rust which might damage bearings and seals when shafts are removed; then remove pinion shafts and side covers as assemblies by bumping threaded end of shafts with a soft hammer or block. With pinion shafts removed,

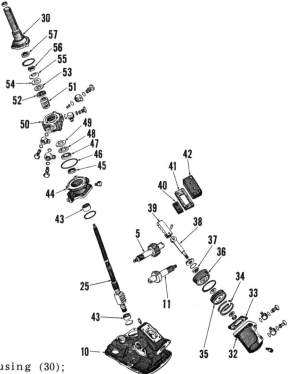

Fig. 11 — Exploded view of power steering gear used on Model MF135. Refer also to Fig. 10.

32. Cylinder
33. Gasket
34. Piston rings
35. Piston
36. Adapter
37. Seal
38. Piston rod
39. Rack
40. Rack guide
41. Shims
42. Guide cover
43. Needle bearing
44. Adapter housing
45. Oil seal
46. "O" ring
47. Bearing race
48. Thrust bearing
49. Bearing race
50. Valve body
51. Valve spool
52. Bearing race
53. Thrust bearing
54. Bearing race
55. Belleville washer
56. Nut
57. Oil seal

unbolt steering shaft housing (30); then withdraw housing and shaft & ball nut assembly (25).

Shaft and ball nut unit (25) is available only as an assembly and component parts are not serviced. If trouble exists, the complete shaft assembly must be renewed, with no attempt made at disassembly.

Shims (S) on adjusting screws (4 & 12) are available in thicknesses of 0.063, 0,065, 0.067 and 0.069. When assembling the steering gear, use a shim of the proper thickness to provide a minimum clearance without binding between adjusting screw head and slot in pinion shaft (5 or 11).

When reassembling the steering gear, install shaft and ball nut assembly (25), using a shim pack (27) of correct thickness to provide a very slight preload to steering shaft bearings. Shims are available in thicknesses of 0.002, 0.005, 0.010 and 0.030 in a shim kit assortment only. Turn steering shaft until lower tooth on ball nut (25) is visible through right side opening of housing and install pinion shaft (11) with solid portion down and first tooth of ball nut meshed in first tooth space of pinion shaft.

Turn steering shaft until center tooth on upper portion of pinion shaft (11) is in line with left side opening in housing and install pinion shaft (5) with center groove mating with upper center tooth on shaft (11). When properly installed, steering wheel should rotate 3½ turns from lock to lock. Complete the assembly by reversing

the disassembly procedure. Fill gear housing with lubricant and adjust backlash as outlined in paragraph 9.

15. **OVERHAUL POWER STEERING GEAR.** Refer to Fig. 11 for exploded view of steering gear. To disassemble the gear, first disconnect the two hoses and external lines from steering valve and cylinder. Remove the steering wheel, steering arms and the cap screws retaining steering gear side covers. Check splined ends of pinion shafts (5 & 11) and remove any paint, burrs or rust which might damage bearings and seals when shafts are removed. Unbolt and remove the power rack guide cover (42), being careful not to lose the shim pack (41). Remove pinion gears (5 & 11) and their respective side covers using a plastic hammer; then unbolt and remove cylinder (32) and steering shaft (25) and associated parts from gear housing (10).

Refer to paragraph 29 for service on the steering cylinder and to paragraph 28 for disassembly procedure of steering shaft and service on steering valve.

Steering shaft (25) is available only as an assembly with the ball nut; if any part is damaged, the complete assembly must be installed, therefore disassembly of ball nut should not be attempted.

When reassembling the steering gear, first reinstall steering shaft assembly and power piston assembly. DO NOT install rack guide (42) at this time. Turn steering shaft until lower tooth on ball nut is aligned with right side cover opening then install pinion gear (11) with first tooth space (next to blank portion) meshed with lower tooth on ball nut. If not previously done, remove side cover from pinion gear (5) by threading adjusting screw out of cover. Turn steering wheel shaft until middle tooth space on upper part of pinion gear (11) is centered in left side cover opening and slide pinion rack (39) until middle tooth space is centered on top portion of side opening. Insert upper pinion gear (5) through left side opening with center tooth of each gear portion meshed with the previously aligned mating tooth spaces as shown in Fig. 12. Complete the assembly by reversing the disassembly procedure, omitting installation of the power piston rack guide until backlash is adjusted as follows:

With both backlash adjusting screws backed out and rack guide removed, turn steering wheel to the center or mid-position. Turn right side adjusting screw until a pull of 1¼-1¾ lbs. (measured at steering wheel rim) is required to turn steering wheel through the mid-position. Turn left side adjusting screw until rolling torque is increased to 2-2½ lbs. when measured in the same manner. Using the removed shim pack or one of equal thickness, reinstall and tighten rack guide (42—Fig. 11); then recheck

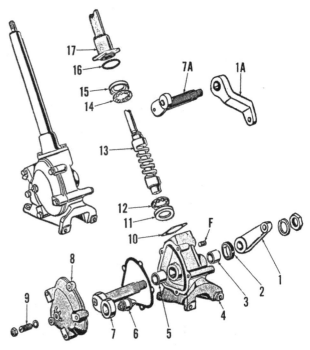

Fig. 14 — Exploded view of steering gear of the type used on Models MF-150 and MF-165.

1. Steering arm (MF150)
1A. Steering arm (MF165)
2. Oil seal
3. Bushing
4. Housing
5. Bushing
6. Stud (MF150)
7. Lever shaft (MF150)
7A. Lever shaft (MF165)
8. Side cover
9. Adjusting screw
10. Shims
11. Bearing cup
12. Bearing
13. Worm shaft
14. Bearing
15. Bearing cup
16. "O" ring
17. Steering column

the rolling torque by turning steering wheel through the mid-position. If shim adjustment is correct, rolling torque should now read ⅛-½ lb. greater than last previous reading. If reading is unchanged, remove rack guide and discard a shim or shims and recheck. If steering gear binds or rolling torque is increased by more than ½ lb., add shims. Shims (41) are available in thicknesses of 0.003, 0.005 and 0.010. Refill steering gear with SAE 90 gear oil until fluid is level with side cover retaining cap screw (L—Fig. 7). Housing may be filled with a suitable pump gun through cap screw hole (L), or by removing breather located on top of gear housing.

Models MF150 - MF165

16. REMOVE AND REINSTALL. To remove the steering gear, remove hood, rear side panels, battery, steering wheel and instrument panel. Remove fuel tank; or block up tank and remove rear support and support frame. Steering gear housing may now be removed; or overhauled without removal from transmission housing. Install by reversing the removal procedure.

17. OVERHAUL. Refer to Fig. 14. To overhaul the steering gear, remove pitman arm (1 or 1A) from pitman shaft (7 or 7A). Remove side cover (8) and withdraw lever shaft and stud assembly (7 or 7A) from housing (4).

Remove steering column (17) and steering shaft (13), saving shim pack (10) for reinstallation when gear is reassembled.

Examine all parts and renew any which are questionable. The roller bearing type lever stud (6) used on MF150 is renewable. Tighten the stud nut to provide a rolling torque of 6-8 inch pounds for the stud bearings. Bushings (3 & 5) for the lever shaft should be reamed after installation, to an inside diameter of 1.6235-1.625, to provide a diametral clearance of 0.0005-0.003 for the lever shaft. Pitman arm (1 or 1A) and shaft are provided with a master spline for correct timing. Arm should point up on Model MF150; down on Model MF165. When installing camshaft (13), vary the thickness of shim pack (10) to remove all end play and provide camshaft bearings with a very slight drag when camshaft is turned in bearings. Shims are available in thicknesses of 0.002, 0.003 and 0.010. Adjust steering gear backlash as outlined in paragraph 11, after unit is reassembled.

18. FRONT STEERING PEDESTAL (Model MF150, Manual Steering).

NOTE: For service on front steering pedestal on power steering models, refer to paragraph 33 or 34.

To remove the upper steering pedestal (10—Fig. 15) and associated parts, first remove the grille, grille pan and left side panel. Loosen the cap screw retaining steering arm to

Fig. 12—Sector gears must be correctly timed during installation. Refer to text.

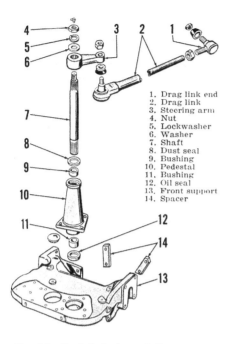

1. Drag link end
2. Drag link
3. Steering arm
4. Nut
5. Lockwasher
6. Washer
7. Shaft
8. Dust seal
9. Bushing
10. Pedestal
11. Bushing
12. Oil seal
13. Front support
14. Spacer

Fig. 15—Exploded view of front support and upper steering pedestal used on Model MF150 with manual steering.

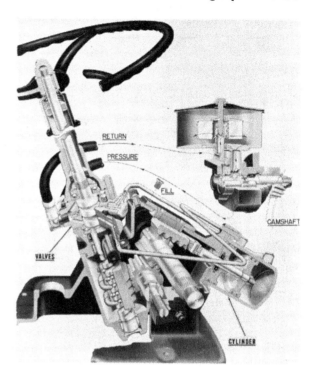

Fig. 17—Cutaway view of power steering system of the type used on Model MF135.

lower end of pedestal shaft (7) and disconnect drag link (2) at forward end. Remove the capscrews retaining pedestal housing (10) to front support (13) and lift off the unit.

Remove nut (4) and disassemble the unit. Examine all parts and renew any which are worn, scored or damaged.

Ream bushings (9 & 11) after installation, to an inside diameter of 1.5005-1.5015. Pedestal shaft (7) should have a diametral clearance of 0.0005-0.002 in bushings. Assemble the unit by reversing the disassembly procedure. Steering arms and shaft (7) are provided with master splines for correct assembly.

TROUBLE SHOOTING
Model MF135

20. On Model MF135, the power steering control valve is actuated by vertical movement of steering wheel shaft and power is applied at the steering gear; refer to Fig. 17. The power steering pump is gear driven from the engine camshaft.

Loss of power assistance could be caused by lack of fluid, a sticking control valve, internal leakage within the cylinder, an improperly adjusted or malfunctioning relief valve or a malfunctioning pump.

Erratic action could be caused by air in the system, a sticking control valve, binding in steering gear or linkage, or the use of incorrect operating fluid.

POWER STEERING SYSTEM

NOTE: The maintenance of absolute cleanliness is of utmost importance in the operation and servicing of the hydraulic power steering system. Of equal importance is the avoidance of nicks and burrs on any of the working parts.

FILLING AND BLEEDING

All Models

19. Fluid capacity is 2/3 qt. for MF135 or MF150 non-diesel; 1⅓ qts. for MF165 non-diesel; 1 qt. for MF135 or MF150 diesel; and 1 2/3 qts. for MF165 diesel. Automatic Transmission fluid, Type A, is the recommended operating fluid for all models. The power steering fluid reservoir is mounted on the gear-driven power steering pump, on left side of engine block on MF165 diesel, right side of block on other models. Fluid should be maintained at level of filler plug on

diesel models and at "FLUID LEVEL" mark on side of reservoir on non-diesels.

To bleed the system, fill reservoir and start the engine, then maintain fluid level at or near full level by adding fluid, as steering wheel is turned to bleed air from the system.

Fig. 18 — Cross sectional view of power steering system of the type used on Model MF150. Model MF165 steering is similar in operation, but differs in appearance and construction.

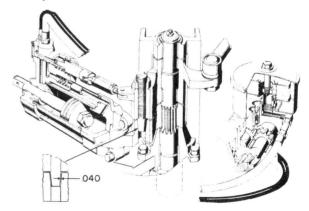

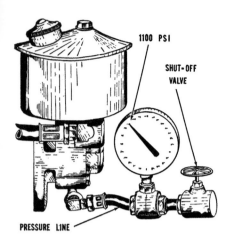

Fig. 19—View of typical power steering pump with pressure gage and shut-off valve installed for checking relief pressure. Refer to text.

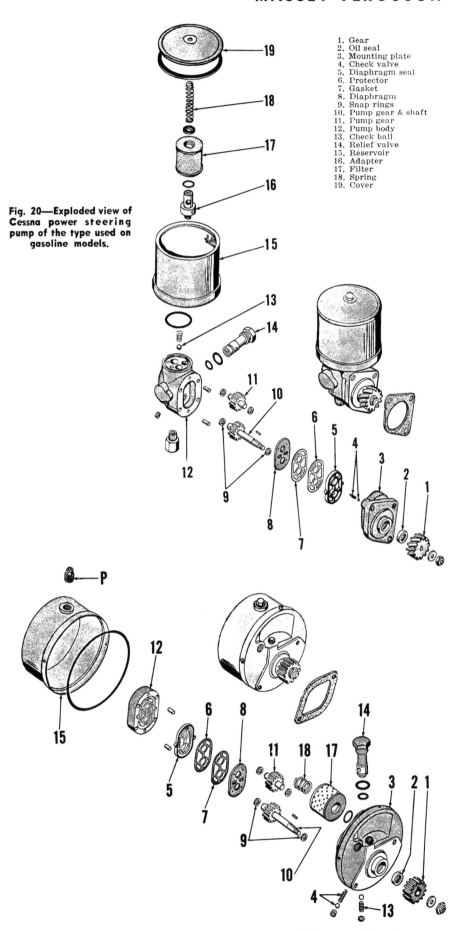

Fig. 20—Exploded view of Cessna power steering pump of the type used on gasoline models.

1. Gear
2. Oil seal
3. Mounting plate
4. Check valve
5. Diaphragm seal
6. Protector
7. Gasket
8. Diaphragm
9. Snap rings
10. Pump gear & shaft
11. Pump gear
12. Pump body
13. Check ball
14. Relief valve
15. Reservoir
16. Adapter
17. Filter
18. Spring
19. Cover

Most causes of failure are accompanied by characteristic noises which will indicate the probable cause. A careful evaluation of the noise and careful inspection will usually help in pinpointing the cause of trouble.

Models MF150 - MF165

21. On these models, the power steering control valve is mounted on the cylinder and attached to the front steering pedestal located inside the radiator grille. Refer to Fig. 18. External adjustments are provided for steering sensitivity and synchronization of the control valve with wheel movement.

Loss of power assistance in one direction only, is most likely caused by improper adjustment of the control valve link.

Loss of power assistance in both directions is most likely caused by improper adjustment of the sensitivity adjusting screw, but could be caused by lack of fluid, a sticking control valve, internal leakage within the cylinder, an improperly adjusted or malfunctioning relief valve or a malfunctioning pump.

Erratic action could be caused by air in the system, a sticking control valve, improper adjustment, wear or binding in steering linkage or the use of incorrect operating fluid.

Shimmy could be caused by an improperly adjusted sensitivity screw or wear in steering linkage.

NOTE: Although the operating principles and general appearance is similar on MF150 and MF165, the action of the valve and cylinder are reversed. On Model MF150,

Fig. 21—Exploded view of Cessna power steering pump of the type used on three cylinder diesel models. Refer to Fig. 20 for parts identification.

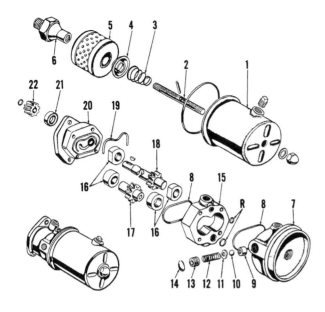

Fig. 22 — Exploded view of Wooster power steering pump of the type used on MF165 diesel.

1. Reservoir
2. Stud
3. Spring
4. Spring seat
5. Filter
6. Retainer
7. Rear cover
8. Seal
9. Valve seat
10. Valve ball
11. Retainer
12. Spring
13. Adjusting screw
14. Plug
15. Pump body
16. Bearings
17. Follow gear
18. Driven gear
19. Seal
20. Front cover
21. Seal
22. Drive gear

the cylinder is extended in making a left-hand turn; on Model MF165, the cylinder retracts. Lengthening the valve link will cause Model MF150 to self-steer to the right; Model MF165 to self-steer to the left. Refer to the proper adjustment paragraphs when making the adjustments.

SYSTEM OPERATING PRESSURE AND RELIEF VALVE

All Models

22. A pressure test of the hydraulic circuit will disclose whether the pump, relief valve or some other unit in the system is malfunctioning. To make the test, proceed as follows:

Connect a pressure test gage and shut-off valve in series with the pump pressure line as shown in Fig. 19. Note that the pressure gage is connected in the circuit between the shut-off valve and the pump. Open the shut-off valve and run engine until operating fluid is warm. Open the throttle and close the shut-off valve only long enough to obtain a reading. Gage reading should be 1100 psi; if it is not, either the pump or relief valve is malfunctioning. On Model MF165 diesel, the relief valve can be adjusted after removing the expansion plug (14—Fig. 22). On all other models, the relief valve cartridge (14—Fig. 20 or 21) is renewable only as a unit, and is not adjustable.

POWER STEERING PUMP

Model MF165 diesel tractor is equipped with a Wooster, gear type pump shown exploded in Fig. 22. All other models use a Cessna gear type pump, refer to Fig. 20

or 21. All pumps are gear driven from the engine camshaft. Refer to the appropriate following paragraphs for overhaul procedures.

Cessna Pump

23. Refer to Fig. 20 or Fig. 21 for exploded view of pump. If pump body or gears are worn or scored, it is recommended that the complete pump be renewed.

When reassembling the pump, make sure that check valve (4) is properly installed in the correct hole. Install diaphragm seal (5) open side down and work the seal firmly in its groove using a blunt tool. Press protecter gasket (6) and plastic gasket (7) into relief in diaphragm seal; then install diaphragm (8) with bronze face toward the gears. Install shaft seal (2) with lip toward inside of pump. After assembly, pump should have a slight amount of drag, but should rotate evenly.

Bleed the system as outlined in paragraph 19 after pump is reinstalled.

Wooster Pump

24. Refer to Fig. 22 for an exploded view of pump. All parts are available individually; however, it is recommended that the pump gears (17 & 18) and four bearings (16) be renewed at the same time if any are damaged because of wear or scoring. If center housing (15) is also worn or scored, it is recommended that the pump be renewed.

To assemble the pump, lubricate the large "O" ring (8) and the two small "O" rings (R) and install them in their grooves; then position center housing (15) on rear body (7) and

secure by temporarily installing two of the assembly bolts. Fit two of the bearings (16) in center housing with relief grooves up (toward gear pocket). Position the gears (17 & 18) in the installed bearings; then install the remaining bearings (16) with oil relief grooves toward gears. Install seal (21) in housing (20) with lip toward the rear, after first coating outside of seal with a sealing compound. Lubricate and install the remaining "O" ring (8) and seal strip (19) in housing (20). Lubricate the lip of oil seal (21). Remove the two previously installed assembly bolts; then carefully slide housing (20) over extended portion of gear shaft (18). install the four assembly bolts and tighten evenly to a torque of 28-32 ft.-lbs.

Fill and bleed the system as outlined in paragraph 19, after pump is installed and check and adjust relief pressure as outlined in paragraph 22.

ADJUSTMENT

Model MF135

25. On Model MF135 tractors, all adjustments of the power steering control valve and cylinder are a necessary part of the overhaul procedure, and no external adjustments are provided. Refer to paragraph 28 for overhaul and adjustment procedure for the control valve; and to paragraph 29 for steering cylinder.

Model MF150

26. Adjustments are provided for synchronization of control valve linkage and for valve sensitivity. Refer to Fig. 23 and proceed as follows:

Open the grille door and remove pin (P) connecting the valve linkage to actuating arm. With engine running at about ½-throttle and wheels in a straight-ahead position, fully loosen the locknut (L) and turn adjusting pin (1) clockwise until it bottoms. Loosen both locknuts (N) and turn linkage nut (2) if necessary, until linkage pin (P) can be inserted through actuating arm and link without moving the control valve. Install and secure the pin (P). With linkage connected, back out the adjusting pin (1) seven full turns.

Check the adjustment by turning steering wheel. If more ease of steering is desired, back out adjusting pin (1) an additional amount. Do not back out pin (1) more than 12 turns.

With linkage properly adjusted, steering effort should be approximately equal for a right-hand or left-hand turn. If tractor steers more easily

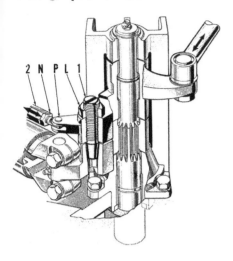

Fig. 23—Cross sectional view of upper pedestal and actuating linkage used on Model MF150. Sensitivity is adjusted by turning the tapered pin (1) in or out of actuating arm. Valve is synchronized by lengthening or shortening link (2). Refer to text for details.

when making a right-hand turn, shorten valve link slightly by turning adjusting sleeve (2). If tractor steers more easily to the left, lengthen the valve link.

Tighten locknuts (L & N) when adjustment is correct, then recheck to be sure adjustment has not changed when locknuts were tightened.

Model MF165

27. Adjustments are provided for valve sensitivity and synchronization of the valve linkage. Refer to Fig. 24 and proceed as follows:

Open the grille door and remove pin (P) which connects valve link (L) to actuating arm (A). With engine running at about ½-throttle and wheels in a straight-ahead position, loosen locknut (1) and turn adjusting

Fig. 24—View of power steering cylinder and actuating linkage used on Model MF-165. Valve is synchronized by shortening or lengthening link (L); sensitivity adjusted by turning the tapered pin (2) in or out of arm (A).

pin (2) clockwise until it bottoms. Loosen the two locknuts (N) and turn adjusting sleeve (S) until pin (P) can be reinserted without moving control valve. Install and secure pin (P); then back out adjusting pin (2) seven full turns.

Check the adjustment by turning steering wheel. Sensitivity can be increased by backing out adjusting pin or decreased by turning in the pin. If tractor steers easier when making a right-hand turn, lengthen valve link (L). If tractor steers more easily to left, shorten link (L).

Tighten the locknuts (1 & N) when adjustment is correct, then recheck the steering response, making adjustments as required.

CONTROL VALVE & CYLINDER
Model MF135

28. **CONTROL VALVE.** For service on the control valve, first disassemble the steering gear as outlined in paragraph 15. Refer to Fig. 25.

Remove the three cap screws securing the valve housing and shaft housing (30) to adapter (44) and lift off shaft housing (30). Clamp steering shaft (25) in a vise and unstake and remove adjusting nut (56). Shaft assembly (25) can now be withdrawn downward out of control valve and adapter (44). Needle bearing (43) can be renewed at this time.

Thrust bearings (48 & 53) are available only as assemblies which includes the respective thrust washers (races). Valve housing (50) and valve spool (51) are available only as a matched pair.

Disassemble and clean the valve and examine for wear, scoring or other damage. The locating groove on inside of valve spool and port markings (PR & RT) serve as match marks when assembling valve and spool. Markings are identified by symbol (L—Fig. 26). Install valve body and spool on steering shaft with match marks (L) toward the top, away from gear housing. Five ball-check type actuating plunger assemblies (P) are used; refer to Fig. 27 for correct assembly and installation of plunger units. Solid plunger (1) goes toward the top, away from gear housing. Install upper thrust bearing, washer (55—Fig. 25) and nut (56). Clamp the steering shaft in a vise and tighten nut (56) to a torque of approximately 30 ft.-lbs.; then back nut off ¼-turn and stake in place, to establish the correct shaft movement to actuate the valve. Assemble the steering gear as outlined in paragraph 15, then bleed the system as in paragraph 19.

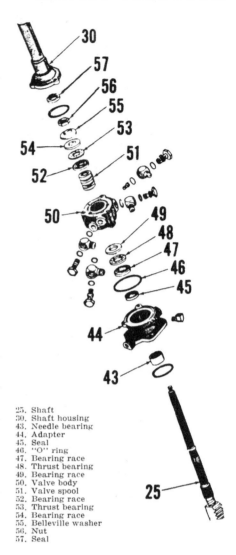

25. Shaft
30. Shaft housing
43. Needle bearing
44. Adapter
45. Seal
46. "O" ring
47. Bearing race
48. Thrust bearing
49. Bearing race
50. Valve body
51. Valve spool
52. Bearing race
53. Thrust bearing
54. Bearing race
55. Belleville washer
56. Nut
57. Seal

Fig. 25—Exploded view of Model MF135 steering shaft and ball nut assembly, steering valve and associated parts.

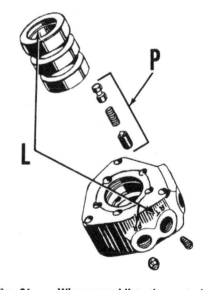

Fig. 26 — When assembling the control valve, make sure that match marks (L) are to the top and that plunger assemblies (P) are installed as shown. Refer to text and to Fig. 27.

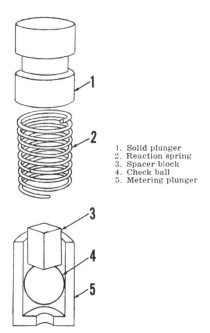

1. Solid plunger
2. Reaction spring
3. Spacer block
4. Check ball
5. Metering plunger

Fig. 27—Cross sectional view of ball check plunger showing correct assembly. Solid plunger (1) goes to top when unit is installed.

29. STEERING CYLINDER. To overhaul the steering cylinder, partially disassemble steering gear as outlined in paragraph 15.

Using a lead or plastic hammer, bump the cylinder off of cylinder end plate and piston. Remove the self-locking nut retaining piston to piston rod and withdraw piston rod from pis-

Fig. 28—Installed view of MF135 steering valve. Correct adjustment is obtained by tightening nut (N); then backing nut off ¼-turn and staking to shaft. Refer to text.

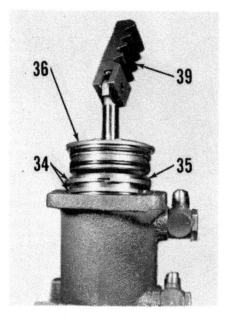

Fig. 29—Partially disassembled view of power steering rack and cylinder used on Model MF135. Refer to text and to Fig. 11.

ton and end plate. Disassemble only far enough to thoroughly clean and examine all parts.

Renew seals and sealing "O" rings and any other parts which are worn, scored or questionable. Reassemble by reversing the disassembly procedure and install and adjust the steering gear as outlined in paragraph 15.

Models MF150 - MF165

30. REMOVE AND REINSTALL. To remove the steering valve and cylinder unit, first remove grille door and disconnect hydraulic pressure and return lines at valve housing. Disconnect control link at actuating arm. Remove piston rod anchor pin and cylinder pivot pin and lift out the complete cylinder and valve unit.

Reinstall by reversing the removal procedure, bleed the system as outlined in paragraph 19 and adjust the valve as in paragraph 26 or 27.

Fig. 30 — Cross sectional view of power steering control valve of the type used on Models MF150 and MF165. Refer to Fig. 31 for parts identification except for the following.

A. Pressure port
B. Return port
C. Cross drilling
D. Pressure passage
E. Pressure passage

31. OVERHAUL VALVE. To disassemble the control valve, refer to Fig. 30 and proceed as follows:

Remove end cover (13). Turn valve spool (17) until cross-drilling (C) is visible through return port (B) and carefully insert a small punch through port (B) and drilling (C). Insert a second punch through pin hole (F) in end of valve rod (24) and unscrew rod end (24) from valve (17) by by turning rod end counter-clockwise. Remove seal (16) and snap ring (22).

NOTE: Snap ring (22) is spiral type in valve used on Model MF150 and Tru-Arc type on Model MF165.

With seal and snap ring removed, push the valve (17), centering washers (20) and centering spring (21) out of valve bore.

Examine valve bore and spool for wear, scoring or other damage, spool and valve housing are available separately and do not require selective fitting. Refer to Fig. 31 for an exploded view of valve and cylinder unit.

When assembling the valve spool, install a new "O" ring (18) in annular groove on valve spool, lubricate the "O" ring with petroleum jelly and insert valve spool carefully from spring end of bore. Install end cap (13) using a new "O" ring (15) to prevent spool from sliding too far into bore. With spring end of valve bore up, install centering washers (20), spring (21) and snap ring (22). Install seal (16). Insert a small punch through return port (B—Fig. 30) and cross drilling (C), make sure "O" ring (23) is properly positioned, and thread rod end (24) into control valve (17). Reinstall cylinder and valve unit as outlined in paragraph 30.

32. OVERHAUL CYLINDER. To overhaul the power steering cylinder, first remove the cylinder as outlined in paragraph 30. Refer to Fig. 31.

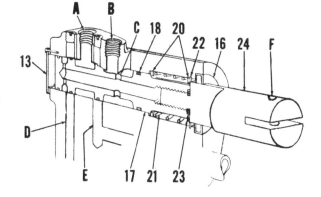

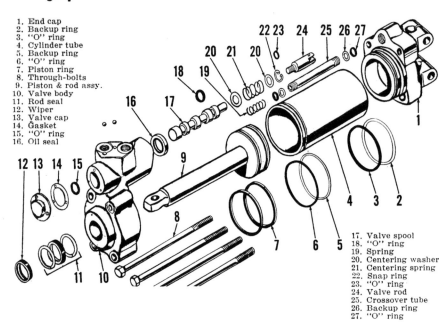

1. End cap
2. Backup ring
3. "O" ring
4. Cylinder tube
5. Backup ring
6. "O" ring
7. Piston ring
8. Through-bolts
9. Piston & rod assy.
10. Valve body
11. Rod seal
12. Wiper
13. Valve cap
14. Gasket
15. "O" ring
16. Oil seal

17. Valve spool
18. "O" ring
19. Spring
20. Centering washer
21. Centering spring
22. Snap ring
23. "O" ring
24. Valve rod
25. Crossover tube
26. Backup ring
27. "O" ring

Fig. 31—Exploded view of power steering cylinder and control valve assembly of the type used on Model MF165. Model MF150 unit is similar and is shown in Fig. 33.

Clamp end cap (1) in a vise with piston rod end up. Remove the four through-bolts (8) and disassemble the cylinder, using Fig. 31 as a guide.

Clean all parts thoroughly and renew any which are scored, worn or damaged. Renew "O" rings, back-up rings and seals whenever cylinder is disassembled. Piston rod seals (11) and wiper (12) can be removed with a sharp pointed tool after piston rod (9) is withdrawn. Be careful not to scratch the bore when removing or installing seals. Backup rings (2, 5 and 26) must be installed away from pressure side of their respective "O" rings; use Fig. 31 as a guide.

Assemble by reversing the disassembly procedure. Tighten the four through-bolts (8) evenly, to a torque of 15-20 ft.-lbs. Reinstall cylinder and valve unit as outlined in paragraph 30.

POWER STEERING PEDESTAL
Model MF150

33. Before removing or disassembling the power steering pedestal,

refer to Fig. 32. Turn wheels to a straight-ahead position, turn adjusting screw (A) clockwise in actuating arm until it bottoms; then punch-mark actuating arm, pedestal bracket and shaft as shown at (M), so parts can be correctly reassembled.

NOTE: There is no master spline on power steering arm (43—Fig. 33) and shaft (37) to assist in assembly.

To remove the pedestal, first remove the grille, lower grille pan and left side panel. Remove power steering cylinder (See paragraph 30) without disconnecting hoses, and lay the cylinder unit out of the way on front support. Disconnect steering drag link at front end.

Loosen the cap screw retaining lower steering arm (42) to steering shaft, unbolt pedestal support (39) from front support and lift the unit from tractor.

Remove snap ring (36) and press steering shaft (37) downward out of pedestal bracket (39), actuating arm (33) and power steering arm (43).

NOTE: DO NOT attempt to remove adjusting pin (31) by backing it out of actuating arm (33) without disassembly of the pedestal. Tapered end of pin is slightly larger than threaded portion, and threads will be damaged if excessive force is applied.

28. Link end
29. Turnbuckle
30. Link end
31. Adjusting pin
32. Nut
33. Actuating arm
34. Bushing
35. Bushing
36. Snap ring
37. Pedestal shaft

38. Bushing
39. Pedestal
40. Bushing
41. Dust seal
42. Steering arm
43. Connecting arm
44. Bushing
45. Pin
46. Bracket
47. Pin

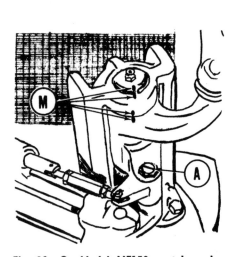

Fig. 32—On Model MF150, match marks (M) should be made on pedestal components before unit is disassembled. Refer to text for procedure.

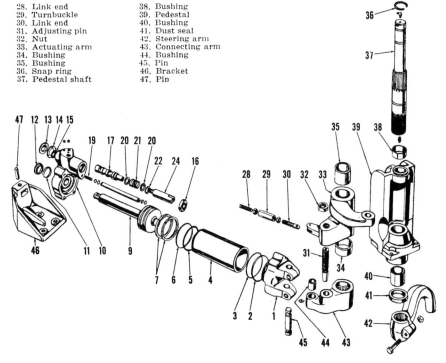

Fig. 33—Exploded view of power steering pedestal, cylinder and associated parts used on Model MF150. Pedestal must be disassembled before adjusting screw (31) can be removed; refer to text. Refer to Fig 31 for parts identification of cylinder and valve unit.

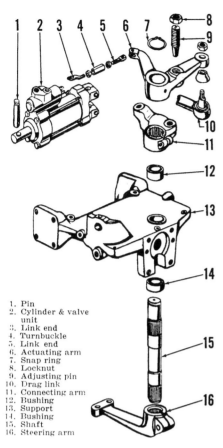

1. Pin
2. Cylinder & valve unit
3. Link end
4. Turnbuckle
5. Link end
6. Actuating arm
7. Snap ring
8. Locknut
9. Adjusting pin
10. Drag link
11. Connecting arm
12. Bushing
13. Support
14. Bushing
15. Shaft
16. Steering arm

Fig. 34—Exploded view of front support, steering pedestal, steering cylinder and associated parts used on Model MF165.

Slide actuating arm (33) and power steering arm (43) out of bracket (39), separate the pieces and remove adjusting pin (31) if necessary, by threading it DOWNWARD out of actuating arm.

If cylinder pivot bushing (44) is to be renewed, align the lubrication hole and press bushing into position flush with both sides of power steering arm (43). Ream the bushing after installation to an inside diameter of 0.7505-0.7508. Lower bushing (40) in pedestal should be align reamed to an inside diameter of 1.8755-1.8765. Upper bushing (38) in pedestal and the two bushings (34 & 35) in actuating arm should be align reamed after installation to an inside diameter of 1.5005-1.5015. All bushings should be installed flush with their respective bores.

Assemble by reversing the disassembly procedure. Tighten the cap screws retaining pedestal to front support to a torque of 47-53 ft.-lbs. and the clamp screw securing lower steering arm (42) to a torque of 33-38 ft.-lbs. Adjust the power steering linkage after installation, as outlined in paragraph 26.

Model MF165

34. Refer to Fig. 34 for an exploded view of power steering pedestal and associated parts. Most service work can be performed without major disassembly of the unit.

Adjusting pin (9) can be threaded upward out of actuating arm (6) if removal is indicated. Actuating arm (6) can be removed after disconnecting control valve link (5) and drag link (10), and removing snap ring (7). Power steering arm (11) is secured to steering shaft (15) by a clamping screw, and can be lifted from shaft after removing actuating arm and loosening the clamp screw.

Bushings (12 & 14) are contained in front support housing (13). Align ream the bushings after installation, to an inside diameter of 1.8755-1.8765.

Renew any damaged, worn or questionable parts, assemble by reversing the disassembly procedure, and adjust the valve linkage as outlined in paragraph 27.

CONTINENTAL NON-DIESEL ENGINE AND COMPONENTS

Model MF135 Special tractors are equipped with a Continental Z-134 engine having a bore of 3.312 inches, a stroke of 3.875 inches and a displacement of 134 cubic inches. Early Model MF135 Deluxe and Model MF150 tractors are equipped with a Continental Z-145 engine having a bore of 3.375 inches, a stroke of 4.062 inches and a displacement of 145 cubic inches. Early Model MF165 tractors are equipped with a Continental G-176 engine have a bore of 3.58 inches, a stroke of 4.38 inches and a displacement of 176 cubic inches. Except for the slight difference in size, all engines have the same general configuration and design features, and overhaul procedures are similar unless otherwise indicated.

R&R ENGINE WITH CLUTCH
All Models

35. To remove the engine and clutch, first drain cooling system and if engine is to be disassembled, drain oil pan. Remove hood. Disconnect drag links and radius rods on Model MF135 and drag link on other models. Disconnect air cleaner hose, radiator hoses; and power steering lines on models so equipped. Support tractor under transmission housing and unbolt front support casting from engine; then roll front axle, support and radiator as an assembly, away from tractor. Shut off the fuel and remove fuel tank. Disconnect heat indicator sending unit, oil gage line, wiring harness, starter cable, tachometer cable and throttle linkage. Support engine in a hoist and unbolt engine from transmission case.

Install by reversing the removal procedure. Bleed the power steering system as outlined in paragraph 19 and adjust the governed speed as in paragraph 150 after tractor is reassembled.

CYLINDER HEAD
All Models

36. REMOVE AND REINSTALL. To remove the cylinder head, drain cooling system, shut off the fuel and remove hood and fuel tank. Disconnect throttle and choke linkage from carburetor, remove air cleaner pipe and exhaust pipe; then unbolt and remove manifold and carburetor assembly. On Model MF135, disconnect throttle control rod at front and rear; on all models, remove the cap screws securing thermostat housing and fuel tank support to cylinder head and move housing out of the way without disconnecting heat indicator sending unit. Remove rocker arm cover and rocker arms assembly and push rods, then unbolt and remove cylinder head from tractor.

Install by reversing the removal procedure. Head gasket is pre-coated and one side is marked "BOTTOM" for proper assembly. Tighten the cylinder head cap screws or stud nuts to a torque of 70-75 ft.-lbs. using the sequence shown in Fig. 36 for models MF135 and MF150; and Fig. 37 for Model MF165. Reinstall push rods.

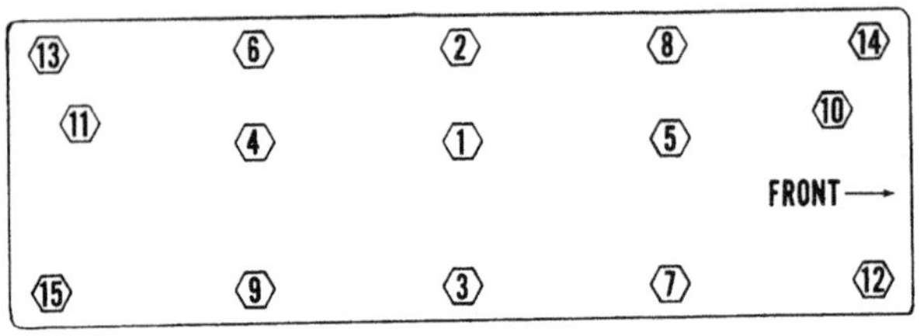

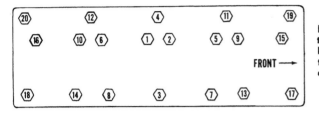

Fig. 36 — Cylinder head tightening sequence for Model MF135 and MF150 gasoline tractors with Continental engine. Recommended tightening torque is 70-75 ft.-lbs.

Fig. 37 — Cylinder head tightening sequence for Model MF165 gasoline tractors with Continental engine. Recommended torque is 70-75 ft.-lbs.

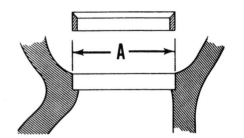

Fig. 39 — Valve seat counterbore (A) should be remachined to accept the 0.010 oversize insert if seats are renewed, refer to text for dimensions.

NOTE: Push rods can be dropped into cylinder block on Models MF135 and MF150. Use care in reinstalling push rods and rocker arms.

Reinstall rocker arms, thermostat housing, manifolds, carburetor, and throttle and governor linkage. Recommended tightening torque for rocker arm support screws is 20-25 ft.-lbs. for Models MF135 and MF150, and 35-40 ft.-lbs. for Model MF165. Tighten manifold stud nuts to a torque of 25-30 ft.-lbs. on Models MF135 and MF-150; on Model MF165, tighten end stud nuts to a torque of 35 ft.-lbs. and remaining stud nuts to a torque of 40-50 ft.-lbs. Adjust valve tappet gap using the procedure outlined in paragraph 39.

On all models, refill radiator, install a nurse tank and run engine until operating temperature is reached; then retorque cylinder head using the sequence shown in Fig. 36 or 37. Readjust valve tappet gap with engine running at slow idle speed. Recommended running tappet clearances (hot) are as follows:

Models MF135 - MF150
All Valves 0.013
Model MF165
Intake 0.016
Exhaust 0.018

37. FUSE PLUG. Most engines are equipped with a thermal type "Fuse"

plug (19—Fig. 38) which serves as a warning device to indicate possible engine damage due to over-heating. The slotted head, ⅛-inch pipe plug threads into the water jacket of the cylinder head. Inner end of plug contains a tin-lead alloy insert, having a melting point of approximately 260° F.

After removing the rocker arm cover and before engine is serviced, remove and inspect the plug. If insert has melted, engine has overheated; check for cracked block or head, warped or damaged gasket surfaces, or other heat damage to engine.

VALVES AND SEATS

All Models

38. Intake valves seat directly in cylinder head and valve stems are equipped with neoprene oil seals. Exhaust valves have renewable seat inserts and stems are equipped with positive type valve rotators (Rotocaps).

Replacement exhaust valve seat inserts are provided in 0.010 oversize only. When renewing the seats, remachine head so that insert bore (A—Fig. 39) measures 1.2595-1.2605 for Models MF135 and MF150; or 1.3535-1.3545 for Model MF165, to provide the recommended 0.003-0.005 interference fit.

Intake valve face and seat angle is 30 degrees. Exhaust valves have a face angle of 44 degrees and seat angle of 45 degrees to provide the recommended 1° interference angle. Desired seat width is 1/16-1/32 inch for all valves. Seats can be narrowed using 15 and 75 degree stones.

39. VALVE TAPPET GAP. The recommended cold tappet gap setting for Model MF165 is 0.018 for intake valves and 0.020 for exhaust; on all other models, recommended setting is

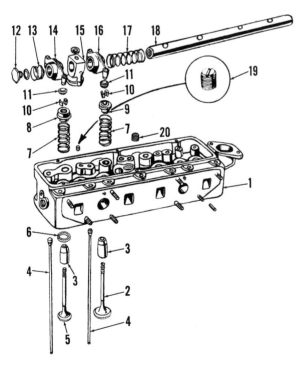

Fig. 38—Exploded view of cylinder head and associated parts used on Continental engines. Fuse plug (19) can be removed to check for evidence of overheating, refer to text.

1. Cylinder head
2. Intake valve
3. Valve guide
4. Push rod
5. Exhaust valve
6. Seat insert
7. Valve spring
8. Valve rotator
9. Retainer
10. Keepers
11. Ball socket
12. End plug
13. Spring
14. Rocker arm
15. Support
16. Rocker arm
17. Spring
18. Rocker shaft
19. Fuse plug

Fig. 40—Align the "DC" flywheel timing mark with timing notch as shown, when adjusting tappet gap as outlined in paragraph 39.

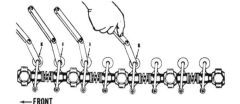

Fig. 41—With "DC" timing mark aligned as shown in Fig. 40 and No. 1 piston on compression stroke, adjust the indicated valves as follows: On Model MF165, adjust No. 1 and No. 3 cylinder exhaust valves (E) to 0.020 and No. 1 and No. 2 intake valves (I) to 0.018. On models MF135 and MF150, adjust all indicated valves to 0.015. Turn crankshaft one complete revolution until timing marks are again aligned as shown in Fig. 40, refer to Fig. 42 and adjust remainder of valves.

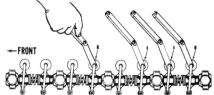

Fig. 42—With "DC" timing mark aligned as shown in Fig. 40 and No. 4 piston on compression stroke, adjust the indicated valves as follows: On Model MF165, adjust No. 2 and No. 4 exhaust valves (E) to 0.020 and No. 3 and No. 4 intake valves (I) to 0.018. On models MF135 and MF-150, adjust all indicated valves to 0.015. Refer also to Fig. 41.

0.015 for all valves. Cold (static) setting of all valves can be made from just two crankshaft positions, using the procedure shown in Figs. 41 and 42, as follows:

Remove timing plug from left side of cylinder block and turn crankshaft until "TDC" timing mark is aligned with timing notch in housing as shown in Fig. 40. Check the rocker arms for No. 1 and No. 4 cylinders. If rocker arms on No. 4 cylinder are tight, No. 1 cylinder is on compression stroke and the tappets indicated in Fig. 41 can be adjusted. If rocker arms on No. 1 cylinder are tight, No. 4 cylinder is on compression stroke and tappets indicated in Fig. 42 can be adjusted. Adjust the four tappets indicated; then turn crankshaft one complete turn and adjust clearance of the remaining four tappets. Recheck the adjustment, if desired, with engine running at slow idle speed.

Recommended tappet clearance with engine at operating temperature is: Intake-0.016, exhaust-0.018 for Model MF165; or 0.013 for all valves on other models. Clearance may be adjusted with engine running at slow idle speed or by following the procedure recommended for initial adjustment.

VALVE GUIDES

All Models

40. The pre-sized intake and exhaust valve guides are interchangeable. Inside diameter of new guides is 0.3157-0.3172 for Models MF135 and MF150; and 0.3420-0.3435 for Model MF165. Inner bore of new guides has a fine, spiral groove or rifling which gives guide an unfinished appearance upon inspection, but guide must not be reamed.

To renew the guides, press old guides downward out of cylinder head using a piloted mandrel and press new guide in from the top, until distance (A—Fig. 43) measured from rocker

arm cover gasket surface to top of guide is 1/16-inch for Model MF165; or 3/32-inch for other models.

Valve stem diameters and clearance limits in guides are as follows:

Models MF135-MF150
Valve Stem Diameter
    Intake 0.3141-0.3149
    Exhaust 0.3124-0.3132
Clearance
    Intake 0.0008-0.0031
    Exhaust 0.0025-0.0048

Model MF165
Valve Stem Diameter
    Intake 0.3406-0.3414
    Exhaust 0.3382-0.3390
Clearance
    Intake 0.0006-0.0029
    Exhaust 0.003-0.0053

VALVE SPRINGS

All Models

41. Intake and exhaust valve springs are interchangeable. Renew any spring which is rusted, discolored or fails to meet the test specifications which follow:

Models MF135-MF150
Free length (approx.) 2-1/16 inches
Lbs. test @ 1 45/64 47-53
Lbs. test @ 1 27/64 96-104

Model MF165
Free length (approx.) 2-5/64 inches
Lbs. test @ 1-21/32 41-47
Lbs. test @ 1-7/32 103-110

VALVE ROTATORS

All Models

42. Normal servicing of the positive type exhaust valve rotators ("Rotocaps") consists of renewing the units. It is important, however, to observe

the valve action after valve is assembled. The valve rotator can be considered satisfactory if the valve turns a slight amount each time the valve opens.

VALVE TAPPETS

All Models

43. The mushroom type tappets (cam followers) operate directly in machined bores of the cylinder block. The 0.5615-0.5620 diameter tappets are furnished in standard size only, and should have a clearance of 0.0005-0.002 in block bores. Tappets can be removed after removing camshaft as outlined in paragraph 47.

Refer to paragraph 39 for valve tappet gap adjustment procedure.

ROCKER ARMS

All Models

44. On model MF165, the rocker shaft is positively positioned by the support bolts, which pass through drilled holes in the shaft and support brackets. Oil holes for the rocker arms must be located on push rod side of engine.

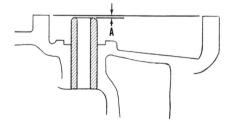

Fig. 43—When renewing valve guides, distance (A) from top of guide to rocker cover gasket surface should measure 1/16-inch on Model MF165 or 3/32-inch on other models.

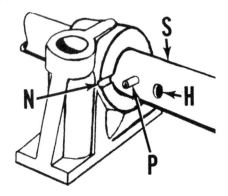

Fig. 44—On Models MF135 and MF150, rocker arm shaft (S) is properly positioned when oil holes (H) are on valve spring side of engine and locating pin (P) fits in notch (N) in front face of center support. Refer to paragraph 44 for proper positioning of shaft on Model MF165.

On all other models, a locating pin (P—Fig. 44) in shaft (S) fits in a notch (N) in front face of center support bracket to locate rocker arm oil holes (H) on valve spring side of engine. All supports are interchangeable and equipped with locating notch (N), however, only the center support serves to locate the shaft.

On Model MF165, the rocker arms should have 0.001-0.0021 clearance on the 0.9671-0.9677 diameter shaft. On other models, rocker shaft diameter is 0.6221-0.623 and recommended clearance of rocker arms is 0.001-0.0029. On all models, the stamped steel rocker arms are right hand and left hand units which are installed in pairs. Bronze bushings in rocker arms are not serviced; if bushing is worn, renew the rocker arm. Refer to paragraph 39 for valve tappet gap adjustment procedure.

TIMING GEAR COVER

All Models

45. **REMOVE AND REINSTALL.** To remove the timing gear cover, drain cooling system and remove hood. Disconnect drag links and radius rods on Model MF135 and drag link on other models. Disconnect air cleaner hoses and radiator hoses; and power steering lines on Models MF150 and MF 165 so equipped. Support tractor under transmission housing, unbolt front axle support from engine and roll axle, support and radiator as an assembly, away from tractor. Remove fan blades and disconnect spring and rod from governor lever. Remove the crankshaft pulley. Remove the cap screws retaining timing gear cover to

engine block and oil pan and remove the cover from its doweled position on engine block.

Oil seal (6—Fig. 45) can be renewed at this time. Install the seal with lip to rear and rear edge flush with seal bore in cover. Refer to paragraph 151 or 152 for procedure for overhaul of governor shaft and associated parts.

Install the cover by reversing the removal procedure. Tighten retaining cap screws to a torque of 25-30 ft. lbs. and adjust governor linkage as in paragraph 150.

TIMING GEARS

All Models

46. Timing gears can be renewed after removing timing gear cover as outlined in paragraph 45. Withdraw governor weight unit from crankshaft on Model MF165 or governor race and shaft assembly from camshaft on other models. Remove the nut retaining camshaft gear to camshaft. On Models MF135 and MF150, lift off governor ball driver assembly. On all models, remove timing gears using a suitable puller. Be careful, when pulling camshaft gear, not to damage governor shaft bore in camshaft on Models MF135 and MF150.

Recommended timing gear backlash is 0.002. Gears are available in standard size; and undersizes or oversizes of 0.001 and 0.002. Gears are marked "S" (Standard), "U" (Undersize), or "O" (Oversize); the "U" or "O" enclosing the number marking "1" or "2" which indicates the exact size.

NOTE: For production reference, a timing gear backlash number is stamped on block front face. When renewing timing gears, any combination of oversize and undersize gears in which the total equals the stamped reference number should give the correct backlash. Refer to (CODE MARKS—Fig. 46).

During installation, mesh the single punch-marked tooth on crankshaft gear with the double punch-marked tooth space on camshaft gear. Heating camshaft gear in oil or in an oven to approximately 300° F. will facilitate gear installation. Remove oil pan and support camshaft in a forward position while gear is being installed, to prevent loosening and leakage of camshaft rear plug.

Tighten the camshaft gear retaining nut to a torque of 70-80 ft.-lbs.

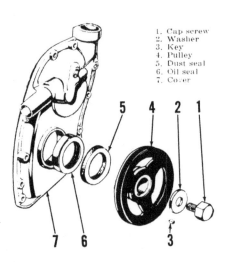

1. Cap screw
2. Washer
3. Key
4. Pulley
5. Dust seal
6. Oil seal
7. Cover

Fig. 45 — Exploded view of timing gear cover and associated parts of the type used on Continental gasoline models.

CAMSHAFT

All Models

47. To remove the camshaft, first remove camshaft timing gear as outlined in paragraph 46. Remove fuel tank, rocker arm cover, rocker arms and shaft assembly; and push rods. Remove the ignition distributor and oil pan. Block up or support the cam followers. Remove the screws securing camshaft thrust plate (18—Fig. 47) to engine block and withdraw camshaft from front of engine.

On model MF165 the camshaft front journal rides in a renewable bushing in engine block and camshaft is further supported by an outboard bear-

Fig. 46—View of timing gear train with timing marks properly aligned. Code marks are used for selection of gears to obtain the correct backlash as outlined in paragraph 46.

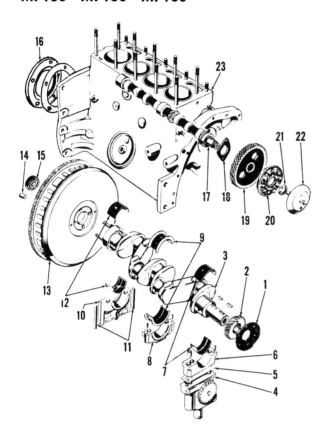

Fig.47 — Exploded view of cylinder block and associated parts of the type used on Models MF135 and MF 150. Model MF-165 is similar except governor weight unit is crankshaft mounted.

1. Oil slinger
2. Crankshaft gear
3. Crankshaft
4. Oil pump
5. Shim pack
6. Main bearing cap
7. Bearing liners
8. Main bearing cap
9. Bearing liners
10. Bearing cap
11. Seal strips
12. Bearing liners
13. Flywheel
14. Cork plugs
15. Pilot bearing
16. Rear oil seal retainer
17. Camshaft
18. Thrust plate
19. Camshaft gear
20. Governor weight unit
21. Nut
22. Governor cup
23. Cylinder block

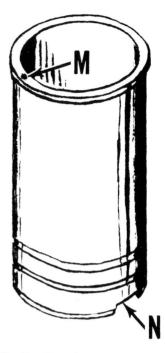

Fig. 48—On all engines except Z134, cut-out notches (N) provide connecting rod clearance and must be installed 90° from crankshaft centerline. Sleeves are correctly installed when locating mark (M) is toward the front.

ing located in timing gear front cover. The two rear journals on Model MF165 and all journals on other models, ride directly in machined bores in engine block. Normal diametral clearance of camshaft journals is 0.0025-0.0045 in block bores; 0.0025-0.0035 for outboard (front) bearing on Model MF165. On Model MF165, renew the camshaft front bushing and/or camshaft if clearance exceeds 0.006. On other models, renew camshaft and/or cylinder block if clearance exceeds 0.007.

Camshaft journal diameters are as follows:

Outboard Journal
(MF165) 0.9965-0.997
Front Journal 1.808-1.809
Center Journal 1.7455-1.7465
Rear Journal 1.683-1.684

Desired camshaft end play of 0.003-0.007 is controlled by thrust plate (18 —Fig. 47) Renew the plate if end play exceeds 0.008.

ROD AND PISTON UNITS

All Models

48. Connecting rod and piston units are removed from above after removing cylinder head and oil pan. Correlation marks on rod and cap should be installed facing camshaft side of engine. Replacement rods are not marked but should be stamped with cylinder number before installation, on side of rod opposite the oil spray hole. Piston skirts are notched at lower edge; notch is to be installed to front of engine when unit is reassembled.

Tighten the connecting rod cap screws to a torque of 40-45 ft.-lbs.

PISTONS, SLEEVES AND RINGS

All Models

49. Pistons are available in standard size only, and are available only in a kit which includes piston, pin, rings and sleeve for one cylinder. Piston is cam ground on all models except MF-165 which uses a cylindrical piston.

If piston and/or sleeve are scored, if piston ring grooves or pin bore are worn or damaged; or if cylinder wall taper exceeds 0.008, renew the piston and sleeve assembly.

Re-ring kits are available. Recommended piston ring end gap is 0.010-0.020 for all rings. Kits are marked for correct piston ring installation. Recommended piston ring side clearance is given in the following table. Renew piston if side clearance exceeds the recommended maximum by 0.0025.

Model MF165

Top compression ring..0.003-0.0045
Second & third compression
rings 0.002-0.0035
Oil control ring 0.002-0.0035

All Other Models

Top compression ring .0.0035-0.005
2nd & 3rd comp. rings 0.0035-0.0055
Oil ring ....Not applicable

Cylinder sleeve on all models except MF135 Special (Z-134 engine) has a cut-out relief clearance for connecting rods as shown at (N—Fig. 48). Relief must be installed at right angles to crankshaft centerline. Most sleeves are punch-marked (M) on top flange 90° from relief slots so correct positioning can be checked from above. Be sure the punch mark is to the front or rear when sleeve is installed.

On all models, use a suitable puller to remove the wet-type cylinder sleeves after rod and piston units are out. Before installing sleeve, clean all cylinder block sealing surfaces. The top of the installed sleeve should extend 0.001-0.004 above top face of cylinder block as shown in Fig. 49. Install the sleeve and measure the stand-out using a straight-edge and

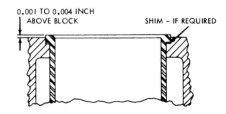

Fig. 49—Cross sectional view of installed cylinder sleeve showing recommended standout.

feeler gage, before installing sealing rings. If standout is excessive, check for foreign matter in sleeve counterbore. Shims are available to correct the trouble, if insufficient standout exists. Excessive or insufficient standout will cause water leaks. After sleeves have been checked, remove the sleeve and install the sealing rings, making sure rings are not twisted. Lubricate the rings and reinstall sleeves, making sure rod clearance reliefs (N—Fig. 48) are properly positioned on models so equipped.

PISTON PINS

All Models

50. The full floating piston pins are retained in piston bosses by snap rings. Piston pins are available in standard size only, for MF165; standard and 0.003 and 0.005 oversize for other models. Recommended clearance for piston pin in both the connecting rod and piston is 0.0001-0.0005 for MF165 and 0.0002-0.0006 for all other models. Specifications are as follows:
Model MF165
　　Pin diameter 1.1250-1.1252
　　Ream bushing to 1.1253-1.1255
All Other Models
　　Pin diameter 0.8591-0.8593
　　Ream bushing to 0.8595-0.8597

CONNECTING RODS AND BEARINGS

All Models

51. Connecting rod bearings are of the precision type renewable from below after removing oil pan. When installing new bearing shells, make sure that the projection engages the milled slot in rod and cap and that rod and cap correlation marks are in register. Replacement rods are not marked and should be stamped with cylinder number on side away from oil squirt hole. Correlation marks should be on camshaft side of block when rods are installed. Bearings are available in undersizes of 0.002, 0.010 and 0.020 as well as standard. Specifications are as follows:

Model MF165
　　Crankpin diameter..2.0615-2.0625
　　Diametral clearance.0.0007-0.0027
　　Wear limit 0.0037
　　Rod side clearance....0.006-0.010
　　Cap screw torque..40-45 ft.-lbs.
All Other Models
　　Crankpin diameter..1.9365-1.0375
　　Diametral clearance.0.0003-0.0023
　　Wear limit 0.0033
　　Rod side clearance ..0.005-0.011
　　Cap screw torque..40-45 ft.-lbs.

CRANKSHAFT AND BEARINGS

All Models

52. The crankshaft is supported in three precision type main bearings renewable from below without removing the crankshaft. The rear main bearing cap (10—Fig. 47) contains packing strips (11) on each side of cap in addition to rear seal (16). To remove the rear main bearing cap and filler block (10), first remove the two cap screws which retain seal (16) to filler block, then unbolt and remove the rear main bearing cap (10).

Bearing inserts are available in undersizes of 0.002, 0.010 and 0.020 as well as standard.

Normal crankshaft end play of 0.004-0.008 is controlled by the flanged center main bearing inserts (9).

To remove the crankshaft, it is necessary to remove engine, clutch, flywheel, rear oil seal, timing gear cover, oil pan and bearing caps.

Specifications are as follows:
Model MF165
　　Journal diameter 2.374-2.375
　　Diametral clearance.0.0005-0.0027
　　Wear limit 0.0037
　　Cap screw torque ..85-95 ft.-lbs.
All Other Models
　Journal diameter 2.249-2.250
　　Diametral clearance.0.0005-0.0027
　　Wear limit 0.0037
　　Cap screw torque ..85-95 ft.-lbs.

CRANKSHAFT REAR OIL SEAL

All Models

53. The crankshaft rear oil seal (16—Fig. 47) is contained in a one-piece retainer and serviced only as an assembly. To renew the seal, first separate engine from transmission case as outlined in paragraph 167 or 168 and remove the flywheel. Remove the oil pan and the two cap screws securing rear seal retainer to main bearing cap; remove the three remaining cap screws and lift off the oil seal and retainer unit.

Apply a light coating of oil to seal lip and gasket sealer to mounting

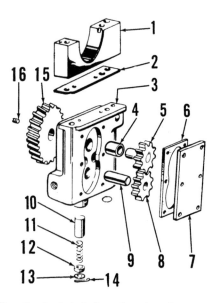

Fig. 50—Exploded view of engine oil pump and associated parts typical of Continental qasoline models.

1. Main bearing cap	10. Relief valve
2. Shim pack	plunger
3. Pump body	11. Spring
4. Bushing	12. Spring seat
5. Driven gear	13. Adjusting shims
6. Gasket	14. Cotter pin
7. Cover	15. Drive gear
8. Follow gear	16. Retaining screw
9. Shaft	

gasket. Position seal retainer with the two threaded holes down. Reinstall cap screws and tighten evenly to a torque of 8-10 ft.-lbs. Complete the assembly by reversing the disassembly procedure.

FLYWHEEL

All Models

54. To remove the flywheel, separate engine from transmission case as outlined in paragraph 167 or 168 and remove the clutch. The starter ring gear can be renewed after removing the flywheel. To install a new ring gear, heat gear evenly to approximately 450° F. and install on flywheel with beveled edge of teeth facing front of engine.

One flywheel mounting stud is off-center so flywheel can only be installed in the correct position. Tighten the flywheel retaining stud nuts to a torque of 70-75 ft.-lbs.

OIL PUMP

All Models

55. The gear type oil pump is mounted on the bottom of front main bearing cap and is gear driven from crankshaft timing gear. Pump is accessible after oil pan is removed.

Shims (2—Fig. 50) control the backlash of oil pump drive gear (15). Recommended backlash is 0.004-

0.008, if backlash is not as specified, vary the thickness of shim pack (2).

Check the pump internal gears (5 & 8) for backlash which should not exceed 0.007. Recommended diametral clearance between gears and pump body is 0.003-0.004; if clearance exceeds 0.005, renew gears and/or body. The pre-sized pump shaft bushing (4) is designed to provide 0.0035-0.0065 diametral clearance for pump shaft. When installing pump cover (7),

tighten the retaining screws to a torque of approximately 8-10 ft.-lbs. Refer to paragraph 56 for data on pressure relief valve.

RELIEF VALVE

All Models

56. The plunger type pressure relief valve (10 through 14—Fig. 50) is

located in oil pump body. Normal operating oil pressure is not less than 7 psi at slow idle speed and 20-30 psi at 1800 rpm. Recommended relief valve setting of 30 psi is obtained by varying the number of adjusting washers (13). The relief valve spring (11) should have a free length of 2 inches and should test approximately 8¼ lbs. when compressed to a working length of 1⅜ inches.

PERKINS NON-DIESEL ENGINE AND COMPONENTS

Late Model MF135 and MF150 tractors are equipped with a Perkins AG3.152 three cylinder gasoline engine having a bore of 3.6 inches, a stroke of 5.0 inches and displacement of 152 cubic inches. Late Model MF165 tractors are equipped with a Perkins AG4.212 four cylinder gasoline engine having a bore of 3.875 inches, stroke of 4.5 inches and displacement of 212.3 inches.

The three cylinder engine is similar in design to its diesel counterpart. Substantial differences exist between the three and four cylinder engines, and between the four cylinder gasoline and diesel engines, but typical Perkins design characteristics are maintained.

R&R ENGINE WITH CLUTCH
All Models

57. To remove the engine and clutch as a unit, first drain cooling system and if engine is to be disassembled, drain oil pan. Remove hood. Disconnect drag links and radius rods on Model MF135 and drag link on other models. Disconnect air cleaner hose, radiator hoses; and power steering lines on models so equipped. Support tractor under transmission housing and unbolt front support casting from engine; then roll front axle, support and radiator as an assembly, away from tractor. Shut off the fuel and remove fuel tank. Disconnect heat indicator sending unit, oil gage line, wiring harness, starter cable, tachometer cable and throttle linkage. Support engine in a hoist and unbolt engine from transmission case.

Install by reversing the removal procedure. Bleed power steering system is outlined in paragraph 19 and adjust governed speed as in paragraph 150 after tractor is reassembled.

CYLINDER HEAD
All Models

58. **REMOVE AND REINSTALL.** To remove the cylinder head, drain cooling system, shut off the fuel and remove hood and fuel tank. Disconnect throttle and choke linkage from carburetor and remove air cleaner pipe and exhaust pipe. On three cylinder engines, disconnect water lines from intake manifold and unbolt and remove intake and exhaust manifold units. On four cylinder units, remove manifold and carburetor assembly. Remove rocker arm cover and rocker arms unit. On four cylinder models remove push rods. On all models, disconnect heat indicator sending unit and oil pressure lines leading to cylinder head, then unbolt and remove cylinder head.

On three cylinder models, mushroom type cam followers operate directly in machined bores in cylinder head and are prevented from falling out as head is removed by the tappet adjusting screw locknuts.

When installing cylinder head, coat both sides of head gasket with a nonhardening sealing compound. Gasket is marked "Top Front" for proper installation. On three cylinder models,

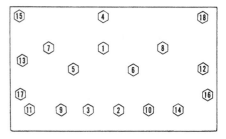

Fig. 51—Cylinder head tightening sequence for Perkins three cylinder gasoline engines. Recommended tightening torque is 55-60 ft.-lbs.

tighten cylinder head stud nuts to a torque of 55-60 ft.-lbs. in the sequence shown in Fig. 51. On Model MF165, tighten cylinder head nuts to a torque of 85-90 lbs. using the sequence shown in Fig. 52. Adjust valve tappet gap as outlined in paragraph 67. Head should be retorqued and valves adjusted after engine is warm.

VALVES AND SEATS
Three Cylinder Models

59. Intake and exhaust valves seat directly in cylinder head. Valve heads and seat locations are numbered consecutively from front to rear. Face and seat angle of new engines is listed as 46 degrees. Use 45° angle for service.

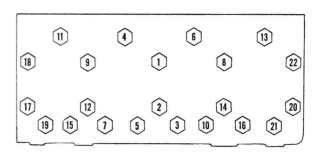

Fig. 52 — Cylinder head tightening sequence for Perkins four cylinder gasoline engine. Recommended tightening torque is 85-90 ft.-lbs.

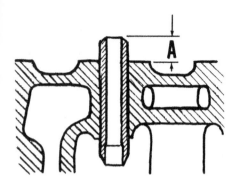

Fig. 53—On three cylinder gasoline models, valve guide is correctly positioned when distance (A) measures 0.37-0.38.

Valve stem diameter is 0.311-0.312 for intake and exhaust. Intake valves are equipped with umbrella type rubber valve stem seals. Intake valves are recessed approximately 0.080 below gasket surface of cylinder head. Exhaust valves are not recessed.

Recommended tappet gap is 0.012 (hot) for intake valves and 0.015 (hot) for exhaust valves. Tappet gap should be adjusted using the procedure outlined in paragraph 67.

Four Cylinder Models

60. Intake and exhaust valves seat directly in cylinder head and production valves and seat locations are numbered consecutively from front to rear. Production face and seat angle is 46 degrees; 45° may be used for service.

Standard valve stem diameter is 0.3725-0.3735 for intake valves and 0.372-0.373 for exhaust valves. Service valves are available in oversizes of 0.003, 0.015 and 0.030 as well as standard. Intake valves are equipped with umbrella type rubber valve stem seals. Intake valves are recessed approximately 0.045 below gasket surface of cylinder head. Exhaust valves are not recessed.

Recommended valve tappet gap is 0.012 (hot) for intake valves and 0.015 (hot) for exhaust. Tappet gap should be adjusted using the procedure outlined in paragraph 67.

VALVE GUIDES

Three Cylinder Models

61. The pre-sized intake and exhaust valve guides are interchangeable and can be pressed from cylinder head if renewal is indicated. Install guide with 45° chamfered edge extending toward spring until distance (A—Fig. 53) measures 0.370-0.380.

Desired diametral clearance for the 0.311-0.312 valve stem in guide is 0.0025-0.0045 for both intake and exhaust valves.

Four Cylinder Models

62. Intake and exhaust valve guides are cast into cylinder head and oversize valve stems are provided for service.

Standard guide bore diameter is 0.3745-0.3755 for both the intake and exhaust valves, with a desired stem to bore clearance of 0.001-0.003 for intake valves and 0.0015-0.0035 for exhaust valves. Oversizes of 0.003, 0.015 and 0.030 are provided in stems of intake and exhaust valves.

VALVE SPRINGS

All Models

63. Springs, retainers and locks are interchangeable for the intake and exhaust valves. Inner and outer springs are used as shown in Fig. 54. Springs may be installed either end up on three cylinder models. Springs for four cylinder engines have close coils which should be installed next to cylinder head. The inner spring (4) has a shorter assembled length than outer spring (5) due to stepped seating washer (3) and retainer (6). Renew springs if they are distorted, discolored or fail to meet the test specifications which follow:

INNER SPRING   Lbs. test @ Inches
Three cylinder models ..... $8@1\frac{3}{16}$
Four cylinder models .....$15 @ 1\frac{9}{16}$
OUTER SPRING
Three cylinder models ....$22 @ 1\frac{1}{2}$
Four cylinder models .....$40 @ 1\frac{25}{32}$

CAM FOLLOWERS

Three Cylinder Models

64. The mushroom type tappets (cam followers) operate directly in machined bores in the cylinder head. The 0.6223-0.6238 diameter cam followers are furnished in standard size only and should have a diametral clearance of 0.001-0.0035 in cylinder head bores.

To remove cam followers after cylinder head is off, first remove adjusting screw and locknut then withdraw cam follower from its bore.

Four Cylinder Models

65. The mushroom type cam followers (tappets) operate directly in machined bores in engine block and can be removed after removing camshaft

as outlined in paragraph 78. The 0.7475-0.7485 diameter cam followers are furnished in standard size only and should have a diametral clearance of 0.0015-0.0037 in block bores. Adjust tappet gap as outlined in paragraph 67.

ROCKER ARMS

All Models

66. The rocker arms and shaft assembly can be removed after removing hood, fuel tank and rocker arm cover. Rocker arms are right hand and left hand units and should be installed on shaft as shown in Fig. 55 or 56. Desired diametral clearance between new rocker arms and new shaft is 0.001-0.0035. Renew shaft and/or rocker arm if clearance is excessive.

The amount of oil circulating to the rocker arms on three cylinder models is regulated by the rotational position of the rocker shaft in support brackets. This position is indicated by a slot on front end of rocker shaft as shown in Fig. 57. When slot is positioned horizontally, the maximum oil circulation is obtained. In production the slot is positioned approximately 30° from vertical and the position indicated by a punch mark on the adjacent support bracket. When reassembling, position rocker shaft slot as indicated by punch mark and check the assembly for proper lubrication. The shaft will not normally need to be moved from the marked position.

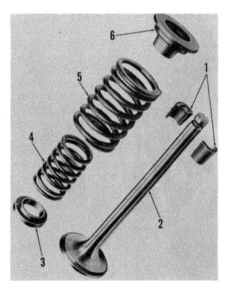

Fig. 54—Disassembled view of valve components, showing correct positioning of springs and retainers.

1. Keepers	4. Inner spring
2. Valve	5. Outer spring
3. Spring seat	6. Retainer

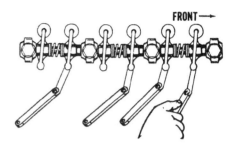

Fig. 55—Assembled view of rocker arm shaft for three cylinder engines

Fig. 56 — Four cylinder gasoline engine rocker shaft showing rocker arms correctly installed. O-ring seal (shown) fits groove in oil supply tube.

Fig. 58—With "TDC-1" timing marks aligned and No. 1 piston on compression stroke, adjust the indicated valves on three cylinder engines. Suggested initial (cold) setting is 0.014 for intake valves and 0.017 for exhaust. Turn crankshaft one revolution and refer to Fig. 59 for remainder of valves.

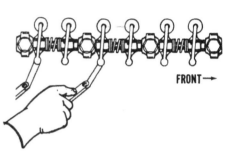

Fig. 59—With "TDC-1" timing marks aligned and No. 1 piston on exhaust stroke, adjust the indicated valves on three cylinder engines. Refer also to Fig. 58.

VALVE TAPPET GAP

All Models

67. The recommended hot tappet gap is 0.012 for intake valves and 0.015 for exhaust valves. Cold (static) settings should be about 0.002 wider than stated hot settings. All valves can be adjusted from just two crankshaft positions using the procedure outlined in this paragraph and illustrated in Figs. 58 through 61.

Remove flywheel timing plug (three cylinder only) and turn crankshaft until "TDC" timing marks are aligned. One three cylinder models, check No. 4 rocker arm counting from front of engine. If valve is open, No. 1 piston is on compression stroke; if valve is closed, No. 1 piston is on exhaust stroke. On four cylinder models, check rocker arms for front and rear cylinders. If rear rocker arms are tight and front rocker arms loose, No. 1 piston is on compression stroke; if front rocker arms are tight and rear

rocker arms loose, No. 4 piston is on compression stroke. With No. 1 piston on compression stroke, adjust the valves indicated in Fig. 58 or 60, turn crankshaft one complete turn until "TDC" marks again align, then adjust remainder of valves. If No. 1 piston is not on compression stroke, adjust the valves shown in Fig. 59 or 61, turn crankshaft one complete turn and adjust remainder of valves.

Fig. 60—On four cylinder models, with "TDC" timing marks aligned and No. 1 piston on compression stroke, adjust the indicated valves. Suggested initial (cold) setting is 0.014 for intake valves and 0.017 for exhaust. Turn crankshaft one revolution and refer to Fig. 61 for remainder of valves.

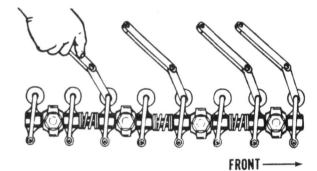

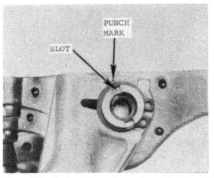

Fig. 57—End view of rocker shaft and support showing indicating slot and punch mark correctly positioned for proper lubrication of rocker arms.

Fig. 61 — With "TDC" timing marks aligned and No. 1 piston on exhaust stroke, adjust the indicated valves. Refer also to Fig. 60. Cylinder head should be re-torqued and valves readjusted when engine is warm.

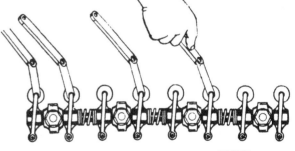

Fig. 62—Front view of three cylinder engine showing parts to be removed before removing timing gear cover.

TIMING GEAR COVER
Three Cylinder Models

68. To remove the timing gear cover, drain cooling system and remove the hood. On Model MF150 with power steering, drain system and remove power steering lines. On all models, disconnect drag links and radius rods if so equipped. Disconnect upper and lower radiator hoses. Support tractor under transmission case, unbolt front support from engine block and roll front support, front axle and radiator as an assembly away from tractor.

Remove crankshaft pulley, water pump and breather pipe. Disconnect carburetor control rod. Remove retaining cap screws and lift off timing gear cover.

Front oil seal bore in cover is not shouldered. Seal should be installed from front until rear edge is approxi-

mately flush with rear of bore and front edge is 0.320-0.330 from front machined face of cover. A Special Tool, MFN 747AA is available to properly position the seal in cover and to center the undoweled cover on timing gear housing. Camshaft thrust spring (Fig. 63) is riveted to inside face of cover and controls camshaft end play.

To remove the governor control cross shaft after cover is off, first unbolt and remove internal lever then remove roll pin and external lever. Press the shaft toward oil filler side of cover to remove the blind needle bearing, then in opposite direction to remove right bearing and seal. Make sure 0.008 clearance is maintenance between shaft shoulder and needle bearing as shown in Fig. 64, when unit is reassembled.

When installing cover, use Special Tool MFN 747AA installed in cover bore to center the cover on shaft. If special tool is not available, use crankshaft pulley as a centering tool.

Four Cylinder Models

69. To remove the timing gear cover, first drain cooling system and remove hood. Disconnect drag link, air cleaner hose, radiator hoses, oil cooler hoses and power steering hoses or lines. Support tractor under transmission housing and unbolt front support casting from engine, then roll front axle, support and radiator as an assembly from tractor.

Remove fan belt, fan blades and ignition coil. Remove the cap screw securing pulley to front of crankshaft and check to see that timing marks are legible (See Fig. 65), then remove pulley. Unbolt and remove timing gear cover.

Crankshaft front oil seal can be renewed at this time. Install seal with sealing lip to rear with front edge of seal recessed 0.380-0.390 into seal bore when measured from front of cover. A special tool (MFN 747B) and spacer (MFN 747C) are available to properly position the seal.

The timing gear cover is not doweled. Special tool MFN 747B can be used as a pilot to properly position to the cover when reinstalling. If tool is not available, reinstall crankshaft pulley to center the seal when reinstalling cover retaining screws. Assemble by reversing the disassembly procedure, making sure pulley timing marks are aligned as shown in Fig. 65. Tighten the pulley retaining capscrew to a torque of 280-300 ft.-lbs.

TIMING GEARS
All Models

70. Refer to Figs. 66 and 67 for views of timing gear trains with cover removed. Before attempting to remove any of the timing gears, first remove fuel tank, rocker arm cover and rocker arms to avoid the possibility of damage to pistons or valve train if camshaft or crankshaft should either one be turned independently of the other.

Because of the odd number of teeth in idler gear, all timing marks will align only once in 18 crankshaft revolutions. To accurately check the timing, turn crankshaft until the proper marks on crankshaft gear and camshaft gear mesh with idler gear; then remove and install idler gear with marks aligned. Governor drive gear does not need to be timed on either engine, but alignment of marks is a

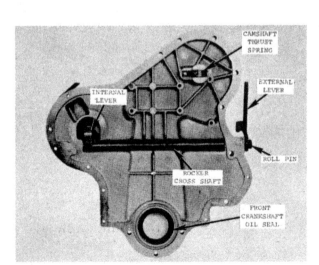

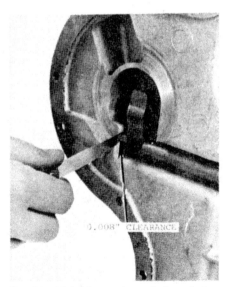

Fig. 63 — Rear (inside) view of three cylinder engine timing gear cover showing oil seal, governor control shaft and camshaft thrust spring.

Fig. 64—Maintain governor shaft end play of 0.008 when installing shaft needle bearings.

Fig. 65—On four cylinder engine, crankshaft pulley alignment marks insure proper positioning of engine timing marks.

convenience for engine assembly. Crankshaft timing gear also drives oil pump on three cylinder models or engine balancer on four cylinder. Balancer must be timed to crankshaft as outlined in paragraph 88.

Service gears are available in standard size only. Refer to the appropriate following paragraphs for renewal of gears, idler shaft or bushings if any of the parts are damaged or if noise is excessive.

71. **IDLER GEAR AND HUB.** Diametral clearance of idler gear on hub should be 0.0012-0.0036 on three cylinder models or 0.0027-0.0047 on four cylinder. Permissible end play is 0.005-0.015 for three cylinder engines or 0.003-0.007 for four cylinder.

Fig. 68—Idler gear removed (three cylinder model) showing gear hub and housing cap screws.

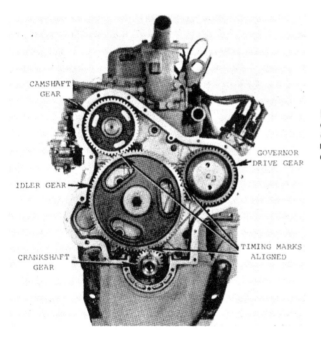

Fig. 66 — Three cylinder engine with timing gear cover removed. Timing marks on idler gear will align only once in 18 crankshaft revolutions.

Idler gear hub is a light press fit in timing gear housing bore. On three cylinder models, hub is further positioned by a locating dowel as shown in Fig. 69. Due to uneven spacing of the three studs on four cylinder models, hub can only be installed in one position. Tighten the hub retaining cap screw on three cylinder models to a torque of 45-50 ft-lbs., or the three screws (four cylinder models) to 20-25 ft-lbs. Measure end play with feeler gauge (Fig. 70) after idler gear is installed.

72. **CAMSHAFT GEAR.** On three cylinder models, camshaft gear is retained by three equally spaced cap screws. To permit installation of the gear for proper engine timing, a letter "D" is stamped on camshaft hub and face of gear as shown in Fig. 71. Align

Fig. 67 — Timing gear train of four cylinder engine. Governor drive gear timing marks align only occasionally but alignment is not necessary for proper engine timing.

Fig. 69—On three cylinder engine, idler gear hub is positioned by a locating dowel as shown.

Fig. 70—Idler gear end play should be 0.005-0.015 for three cylinder engines or 0.003-0.007 as shown for four cylinder units.

the stamped letters when gear is installed. Tighten retaining cap screws to 20 ft.-lbs.

On four cylinder models, camshaft gear is keyed to shaft and retained by a cap screw. Camshaft gear is a transition fit (0.001 tight to 0.001 clearance) and threaded holes are provided for pulling gear. Make sure timing marks are aligned and tighten retaining cap screw to 45-50 ft.-lbs.

73. GOVERNOR GEAR. On all models, governor gear also drives the distributor. Gears contain timing marks which are convenient for assembly but not essential for engine timing, as timing can be accomplished externally when distributor is installed.

On three cylinder models, the nut which retains gear and ball race to distributor drive shaft is left-hand thread, and so marked as shown in Fig. 72. The gear is a transition fit (0.0007 tight to 0.0015 clearance) on distributor drive shaft and keyed in position. Distributor drive shaft is retained in bore by a thrust plate at

Fig. 72—On three cylinder models, governor gear shaft nut is left-hand thread as shown.

rear of shaft and can be removed as outlined in paragraph 79. When installing drive gear and governor ball race unit, tighten left-hand nut to a torque of 25-30 ft.-lbs.

On four cylinder models, governor gear is a slip fit (0.0003-0.0012 clearance) on shaft. Shaft contains governor weight unit as well as distributor drive and unit can be removed from rear as outlined in paragraph 80 after gear is off. When installing gear, tighten retaining nut to a torque of 25-35 ft.-lbs.

74. CRANKSHAFT GEAR. Crankshaft gear is keyed to shaft and is a transition fit (0.001 tight to 0.001 clearance) on shaft. It is usually possible to remove the gear using two small pry bars to move the gear forward. Oil pump or engine balancer will need to be removed if a puller is required.

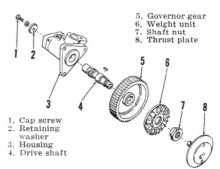

5. Governor gear
6. Weight unit
7. Shaft nut
8. Thrust plate

1. Cap screw
2. Retaining washer
3. Housing
4. Drive shaft

Fig. 74—Exploded view of distributor drive shaft, housing, governor gear and associated parts used on three cylinder models.

75. TIMING THE GEARS. To install and time the gears, first install camshaft and crankshaft gears as outlined in the appropriate preceding paragraphs with timing marks to front. Turn the shafts until the appropriate timing marks point toward idler gear hub, then install idler gear with marks aligned as shown in Fig. 66 or 67. The timing marks on governor drive gear should be aligned for convenience when all gears are removed, but proper alignment is not necessary for ignition timing. Secure idler gear as outlined in paragraph 71.

TIMING GEAR HOUSING
All Models

76. To remove the timing gear housing, first remove timing gears as outlined in paragraphs 68 through 74 and the distributor and drive unit as in paragraphs 79 or 80. On three cylinder models, remove camshaft, oil pan and front bridge piece. On four cylinder models, timing gear housing must be removed before camshaft can be withdrawn.

Fig. 71—Camshaft gear is correctly installed on three cylinder engine when stamped "D's" are aligned as shown.

Fig. 73—The four cylinder engine camshaft is retained by thrust washer which can be removed only after removing timing gear housing.

Fig. 75—Lift out distributor unit.

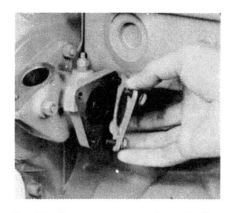

Fig. 76—Remove rear cover from distributor drive shaft housing.

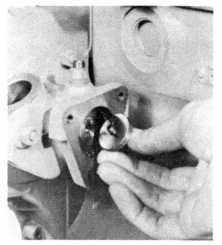

Fig. 77—Remove retaining thrust plate.

Fig. 78—Distributor drive shaft can be withdrawn from front after retaining thrust plate is removed.

Fig. 79—Unbolt and remove distributor drive shaft housing.

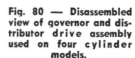

Fig. 80 — Disassembled view of governor and distributor drive assembly used on four cylinder models.

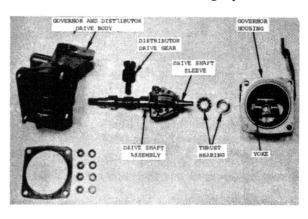

On all models, remove the cap screws retaining timing gear case to engine block (and oil pan on four cylinder units) and lift off timing gear housing. Install by reversing the removal procedure.

CAMSHAFT

Three Cylinder Models

77. To remove the camshaft, first remove timing gear cover as outlined in paragraph 68. Remove fuel tank, rocker arm cover and rocker arms assembly. Secure valve tappets (cam followers) in their uppermost position and withdraw camshaft and gear as a unit.

Camshaft end float is controlled by thrust spring (Fig. 63). The three camshaft journals have a recommended diametral clearance of 0.004 -0.008 in block bores. Journal diameters are as follows:

Front 1.869-1.870
Center 1.859-1.860
Rear 1.839-1.840

Four Cylinder Models

78. To remove the camshaft, first remove timing gears, distributor drive (governor) and timing gear housing as outlined in paragraphs 69 through 76. Secure cam followers (tappets) in the uppermost position and lift off thrust washer (Fig. 73), then withdraw camshaft from block bores.

Thrust washer (Fig. 73) retains camshaft and controls end play. Thrust washer thickness is 0.216-0.218; check thrust washer for correct thickness and for wear or scoring. Recommended camshaft end play is 0.004-0.016 and diametral clearance of camshaft journals in their bores is 0.0025-0.0053. Camshaft journal diameters are as follows:

Front 1.9965-1.9975
Center 1.9865-1.9875
Rear 1.9665-1.9675

DISTRIBUTOR DRIVE SHAFT

Three Cylinder Models

79. Fig. 74 shows an exploded view of distributor drive shaft and associated parts. To overhaul the unit, turn crankshaft until No. 1 piston is coming up on compression stroke and 12° timing mark is aligned with the dot on flywheel housing at timing window. Remove timing gear cover as outlined in paragraph 68 and governor gear as in paragraph 73. Remove distributor as shown in Fig. 75, rear cover (Fig. 76), thrust washer (Fig. 77), drive shaft (Fig. 78) and housing (Fig. 79).

Distributor drive shaft end play is 0.004-0.010 and thrust is always to the rear because of governor action. Diametral clearance of shaft in housing bore is 0.001-0.0025.

Assemble by reversing the disassembly procedure. With No. 1 piston at 12° BTDC on compression stroke, install distributor with rotor pointing toward engine block. Turn distributor body until points just begin to open and install holddown clamp. Complete the assembly by reversing the disassembly procedure. Check timing if necessary, as outlined in paragraph 158.

Four Cylinder Models

80. A disassembled view of distributor drive unit and governor assembly is shown in Fig. 80. To remove or disassemble the unit, first remove timing gear cover as outlined in paragraph 69 and governor gear as in paragraph 73. Remove the distributor as shown in Fig. 81 and governor shaft thrust plate as shown in Fig. 82. Disconnect governor and throttle linkage and unbolt and remove housings as shown in Fig. 83.

Unbolt and remove governor housing from drive body. Withdraw thrust bearing and distributor drive gear then remove drive shaft assembly and governor weight unit.

Distributor drive shaft end play is 0.004-0.008 and thrust is forward, be-

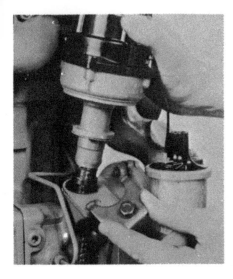

Fig. 81—Remove distributor clamp and lift out distributor assembly.

cause of governor action. Diametral clearance of shaft in housing bore is 0.001-0.0025.

Assemble by reversing the disassembly procedure. Turn crankshaft until No. 1 piston is coming up on compression stroke and 12° timing mark on crankshaft pulley is aligned, then install distributor drive gear with drive slot perpendicular to crankshaft and offset to rear as shown in Fig. 84. Install distributor and turn distributor body until points just begin to open, then install and tighten clamp. Adjust governed speed as outlined in paragraph 150 and timing as in paragraph 158 if adjustment is necessary.

ROD AND PISTON UNITS

All Models

81. Connecting rod and piston units are removed from above after removing cylinder head, oil pan and rod bearing caps (engine balancer on four

Fig. 82—Removing governor shaft thrust plate.

cylinder models). Cylinder numbers are stamped on the connecting rod and cap. When installing rod and piston units, make certain the correlation numbers are in register and face away from camshaft side of engine. Tighten connecting rod nuts to a torque of 40-45 ft.-lbs. on three cylinder engines or 65-70 ft.-lbs. on four cylinder.

PISTONS, SLEEVES AND RINGS

All Models

82. The aluminum alloy pistons have a combustion chamber cavity cast into piston crown and an "F" marking on top of piston to indicate front. Pistons are available in standard size only.

Each piston is fitted with a plain faced chrome top ring which is installed either side up. Four cylinder models use a plain faced second compression ring which may be installed either side up. The internally chamfered second compression ring on three cylinder models or third ring on four cylinder is marked "BTM" (bottom) for correct installation. A chrome railed, segmented oil control ring is used on all engines.

Recommended side clearance for compression rings of all engines is 0.002-0.004. End gap for compression rings should be as follows: All rings, three cylinder engines, 0.011-0.016; top rings, four cylinder engines, 0.016-0.021; other rings, four cylinder engines, 0.012-0.017.

Production sleeves are a tight press fit in cylinder blocks and are finished after installation. Service sleeves are pre-finished and are a transition fit. Sleeves should not be bored and oversize pistons are not available. When installing new sleeves, make sure sleeves and bores are absolutely clean and dry, then chill the sleeves and press fully into place by hand. On three cylinder engine a properly installed sleeve should be flush to 0.004 below gasket surface of engine block.

Fig. 83—Removing governor housing assembly.

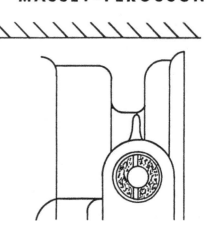

Fig. 84—With No. 1 piston at 12° BTDC on compression stroke, distributor drive slot should be perpendicular to crankshaft and offset to rear as shown. Insert a screwdriver in drive slot and turn counter-clockwise to remove

On four cylinder engine, sleeve should extend 0.030-0.035 above gasket face of block.

PISTON PINS

All Models

83. The full floating piston pins are retained in piston bosses by snap rings and are available in standard size only. The renewable connecting rod bushing must be final sized after installation to provide a diametral clearance of 0.0005-0.0015 for the pin. Be sure the pre-drilled oil hole in bushing is properly aligned with hole in top of connecting rod and install bushing from chamfered side of bore. Piston pin should be a thumb press fit in piston after piston is heated to 160° F. Piston pin diameter is 1.2495-1.250 for three cylinder engine and 0.9998-1.0000 for four cylinder.

CONNECTING RODS AND BEARINGS

All Models

84. Connecting rod bearings are precision type, renewable from below after removing oil pan, balancer unit (four cylinder) and rod bearing caps. When renewing bearing shells, be sure the projection engages the milled slot in rod and cap and that correlation marks are in register and face away from camshaft side of engine.

On three cylinder engines, connecting rod bearings should have a diametral clearance of 0.0025-0.004 on the 2.2485-2.249 diameter crankpin. Bearings are available in undersizes of 0.010, 0.020 and 0.030 as well as standard. Recommended connecting rod side clearance is 0.010-0.015. Tighten the self-locking connecting rod nuts to a torque of 45-50 ft.-lbs.

Fig. 85—Renewable thrust washers (B & C) control crankshaft end play. Main bearing caps are positively located by ring dowels (A).

On four cylinder models, connecting rod bearings should have a diametral clearance of 0.0015-0.003 on the 2.499-2.4995 diameter crankpin. Recommended connecting rod side clearance is 0.010-0.015. Renew the self-locking connecting rod nuts and tighten to a torque of 65-70 ft.-lbs.

CRANKSHAFT AND BEARINGS
Three Cylinder Models

85. The crankshaft is supported in four precision type main bearings. To remove the rear main bearing cap, it is first necessary to remove the engine, clutch, flywheel, flywheel adapter and rear oil seal. Other main bearing caps can be removed after removing oil pan.

Upper and lower bearing liners are interchangeable on all except the front main bearing. When renewing front bearing, make sure the correct half is installed in engine block. Main bearing caps are numbered from front to rear and must be installed with block serial number reading from same direction on block and caps.

Crankshaft end play is controlled by renewable thrust washers (B & C —Fig. 85) at front and rear of rear main bearing. The cap half of thrust washer is prevented from turning by the tab which fits in a machined notch in cap. Block half of washer can be rolled from position when bearing cap is removed. Recommended crankshaft end play is 0.002-0.015. Thrust washers are 0.121-0.123 in thickness and 0.007 oversize washers are available as well as standard.

Bearing inserts are available in standard size and undersizes of 0.010, 0.020 and 0.030. Recommended tightening torque of main bearing capscrews is 110-115 ft.-lbs. Check crankshaft journals against the values which follow:

Main journal diameter . . 2.7485-2.749
Crankpin diameter 2.2485-2.249
Diametral Clearance:

    Main bearings 0.003-0.005

    Crankpin bearings 0.0025-0.004

Four Cylinder Models

86. The crankshaft is supported in five precision type main bearings. To remove the rear main bearing cap it

is first necessary to remove engine, clutch, flywheel and rear oil seal. All other main bearing caps can be removed after removing oil pan and engine balancer.

Upper and lower main bearing inserts are not interchangeable. The upper (block) half is slotted to provide pressure lubrication to crankshaft and connecting rods. Inserts are interchangeable in pairs for all journals except center main bearing. The center main bearing journal controls crankshaft end thrust and renewable thrust washers are installed at front and rear of cap and block bearing bore. Lower half of insert is anchored by a tab to bearing cap; upper half can be renewed after cap is removed.

Bearing inserts are available in undersizes of 0.010, 0.020 and 0.030 as well as standard. Standard thrust washers are 0.089-0.091 in thickness and oversizes of 0.0075 are available as well as standard thickness. Oversizes may be installed in pairs at front or rear of crankshaft; or at front and rear, if required, to correct for excessive end play. When renewing rear main bearing, refer to paragraph 90 for installation of rear seal and oil pan bridge piece. Recommended tightening torque for main bearing cap screws is 145-150 ft.-lbs. Check crankshaft journals against the values which follow:

Main journal diameter . . 2.9985-2.999
Crankpin diameter 2.499-2.4995
Diametral Clearance:

    Main bearings 0.0025-0.0045

    Crankpin bearings 0.0015-0.003

ENGINE BALANCER
Four Cylinder Models

87. OPERATION. The Lanchester type engine balancer consists of two unbalanced shafts which rotate in opposite directions at twice crankshaft speed. The inertia of the shaft weights is timed to cancel out natural engine vibration, thus producing a smoother running engine. The balancer is correctly timed when the balance weights are at their lowest point when pistons are at TDC and BDC of their stroke.

The balancer unit is driven by the crankshaft timing gear through an idler gear attached to balancer frame. The engine oil pump is mounted at rear of balancer frame and driven by the balancer shaft. Refer to Figs. 86 through 88.

88. REMOVE AND REINSTALL. The balancer assembly can be removed after removing the oil pan and

Fig. 86 — Installed view of engine balancer with timing marks aligned. Refer to paragraph 88 for installation procedure.

mounting cap screws. Engine oil is pressure fed through balancer frame and cylinder block. Balancer frame bearings are also pressure fed. Refer to Fig. 86 for an installed view of balancer unit.

When installing balancer with engine in tractor, timing marks will be difficult to observe without removing timing gear cover. The balancer assembly can be safely installed as follows:

Turn crankshaft until No. 1 and No. 4 pistons are at the exact bottom of their stroke. Remove balancer idler gear (5—Fig. 88) if necessary, and align the single punch-marked tooth of idler gear between the two marked teeth on weight drive shaft as shown at (B—Fig. 86). Install balancer frame with balance weights hanging normally. If carefully installed, timing will be correct. If engine is mounted in stand or tractor front end is removed, timing marks can be observed by removing timing gear cover.

NOTE: Balancer can be safely installed with No. 1 & 4 pistons at either TDC or BDC. BDC is selected because interference between connecting rod and balance weights can give warning if unit is badly out of time. Also, alignment of timing marks is not essential but is a convenience if someone else later disassembles the engine.

With balancer correctly installed, tighten the retaining cap screws to a torque of 32-36 ft.-lbs. and complete the assembly by reversing the removal procedure.

89. **OVERHAUL.** Refer to Fig. 88 for an exploded view of balancer frame and associated parts. To disassemble the removed balancer unit, unbolt and remove oil pump housing

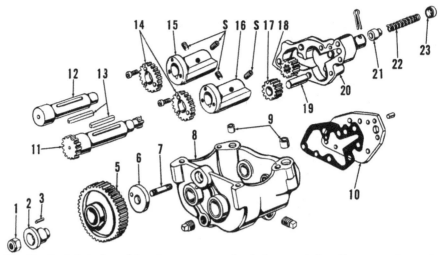

Fig. 88—Exploded view of Lanchester type engine balancer, engine oil pump and associated parts used on four cylinder models.

1. Locknut	7. Stud	13. Key	19. Shaft
2. Hub	8. Frame	14. Gear	20. Pump body
3. Dowel	9. Ring dowels	15. Balance weight	21. Valve piston
5. Idler gear	10. Plate	16. Balance weight	22. Valve spring
6. Washer	11. Drive shaft	17. Pump gear	23. Cap
	12. Driven shaft	18. Pump gear	S. Set screws

(20) and associated parts; and idler gear (5) and associated parts. Set screws (S) retaining balance weights (15 & 16) are installed using Grade "A" (Red) LOCTITE. Loosen the set screws, then push balance shafts (11 & 12) forward out of frame and weights. NOTE: Use care when removing shafts, not to allow keys (13) to damage frame bushings as bushings are not available as a service item.

Recommended diametral clearance for shafts (11 & 12) in frame bushings is 0.002-0.0045 for front bushings or 0.002-0.0035 for rear. Shaft diameters are 1.2425-1.249 at front journal and 0.9987-0.0002 at rear. When assembling the balancer, use Grade "A" (Red) LOCTITE for installing the screws retaing gears (14) to balance weights (15 & 16) and the set screws

(S) retaining balance weights to shafts. Also make sure flat surfaces of weights are aligned when installed, as shown in Fig. 87.

Recommended diametral clearance for idler gear (5—Fig. 88) on hub (2) is 0.001-0.0032. End play should be 0.008-0.014. Bushing for idler gear is not available as a service item.

Refer to paragraph 95 for overhaul of engine oil pump and to paragraph 88 for installation of balancer assembly.

CRANKSHAFT REAR OIL SEAL
All Models

90. The asbestos rope type rear oil seal is contained in a two-piece seal retainer attached to rear face of engine block as shown in Fig. 89. The seal retainer can be removed after

Fig. 87—Assembled view of removed engine balancer and oil pump unit. Refer to Fig. 88 for exploded view.

Fig. 89—Rear view of engine block showing oil seal retainer installed.

removing flywheel and adapter plate on three cylinder models; or flywheel on four cylinder units.

The rope type crankshaft seal is precision cut to length, and must be installed in retainer halves with

Fig. 90—Use a round bar to bed the asbestos rope seal in retainer half. Refer to text for details.

Fig. 91—Cylinder block bridge is equipped with end seals as shown.

0.010-0.020 of seal ends protruding from each end of retainer groove. Do not trim the seal. To install the seal, clamp each retainer half in a vise as shown in Fig. 90. Make sure seal groove is clean. Start each end in groove with the specified amount of seal protruding. Allow seal rope to buckle in the center until about an inch of each end is bedded in groove, work center of seal into position, then roll with a round bar as shown. Repeat the process with the other half of seal retainer.

When installing cylinder block bridge piece on four cylinder engines, insert end seals as shown in Fig. 91. Use a straight edge as shown in Fig. 92 to make sure bridge piece is flush with rear face of cylinder block.

On all models, coat both sides of retainer gasket and end joints of retainer halves with a suitable gasket cement. Coat surface of rope seal with graphite grease. Install retainer halves and cap screws loosely and tighten clamp screws thoroughly before tightening the retaining cap screws.

FLYWHEEL
All Models

91. To remove the flywheel, first separate engine from transmission housing and remove the clutch. Flywheel is secured to crankshaft flange by six evenly spaced cap screws. To properly time the flywheel, align the seventh (unused) hole in flywheel with untapped hole in crankshaft flange. Tighten the flywheel retaining cap screws to a torque of 75-80 ft.-lbs. on all models.

ENGINE ADAPTER PLATE
Three Cylinder Models

92. The engine flywheel is housed in a cast iron adapter plate which is located to engine block by two dowels and secured to block rear face by six cap screws. To obtain access to rear main bearing, crankshaft or rear oil seal it is first necessary to remove the adapter plate. After flywheel is off, adapter plate may be removed by removing the six retaining cap screws and tapping the plate from its doweled position on cylinder block.

OIL PAN
All Models

93. The heavy cast iron oil pan serves as the tractor frame and attaching point for tractor front support. To remove the oil pan, first drain cooling system and oil pan. Remove hood and side panels. On Model MF135, disconnect drag links and radius rods. On other models disconnect drag link. If so equipped, drain and disconnect power steering pump.

On all models, support tractor underneath transmission housing, disconnect radiator hoses, unbolt front support from oil pan and roll front support, front axle and radiator as an assembly from tractor.

Place a rolling floor jack under oil pan, then unbolt and lower oil pan away from engine block. Install by reversing the removal procedure.

OIL PUMP
Three Cylinder Models

94. The rotary type oil pump is mounted in front main bearing cap and driven from the crankshaft timing gear through an idler as shown in Fig. 93.

Fig. 92—Use a straight edge to align the cylinder block bridge.

Fig. 93—On three cylinder models, oil pump mounts on front main bearing cap as shown.

Fig. 94—Engine oil pump can be unbolted from front main bearing cap after removing idler gear.

Fig. 95—Check sealing O-ring for leakage or damage.

To remove the oil pump, first remove oil pan as outlined in paragraph 93. Remove the small lower section of timing gear housing which extends below crankshaft and seals front of oil pan. Pump can be removed as an assembly with front bearing cap or detached from cap as shown in Fig. 94 after removing idler gear.

Check rotors for scoring or excessive wear. Check O-ring (Fig. 95) for leakage or damage. If pump components are excessively worn or otherwise damaged, renew the pump. Component parts are not available through parts stock.

Four Cylinder Models

95. The gear type oil pump is mounted on engine balancer frame and driven by balancer shaft as shown in Fig. 88. Oil pump can be removed after removing oil pan. The thickness of oil pump gears (17 & 18) should be 0.001 greater to 0.004 less than gear pocket depth in pump body (20). Radial clearance of gears in body bores should be 0.0003-0.012. Examine body gears and plate for wear or scoring and renew any parts which are questionable. Refer to paragraph 96 for relief valve.

RELIEF VALVE

All Models

96. The plunger type relief valve is located in oil pump body on all models. Oil pressure should be 30-60 psi at full engine speed with engine at normal operating temperature. Valve spring is retained by a cap and cotter pin.

DIESEL ENGINE AND COMPONENTS

Perkins diesel engines are used. All engines have a bore of 3.6 inches and a stroke of 5.0 inches. Models MF135 and MF150 use a three cylinder engine with a piston displacement of 152 cubic inches. Model MF165 uses a four cylinder engine having a piston displacement of 203.5 cubic inches. All engines have thin-walled cylinder sleeves and a direct injection system. All engines are similar in design and many of the parts are interchangeable.

R&R ENGINE WITH CLUTCH

All Models

97. To remove the engine and clutch as a unit, first drain the cooling system and if engine is to be disassembled, drain oil pan. Remove hood. Disconnect drag links and radius rods on Model MF135 and drag link on other models. Disconnect air cleaner hose, radiator hoses; and power steering lines on models so equipped.

Support tractor under transmission housing and unbolt front support casting from engine; then roll front axle, support and radiator as an assembly, away from tractor.

Shut off fuel and remove fuel tank. Unbolt and remove starting motor and disconnect wiring harness from generator. Disconnect heat indicator sending unit, oil gage line and tachometer cable. Disconnect fuel pressure

and return lines and throttle linkage from injection pump. Support engine in a hoist and unbolt engine from transmission case.

Install by reversing the removal procedure. Bleed the power steering system as outlined in paragraph 19 and fuel system as in paragraph 136 after engine is installed.

CYLINDER HEAD

All Models

98. **REMOVE AND REINSTALL.** To remove the cylinder head, first remove hood and side panels. Shut off the fuel and remove fuel tank. Drain cooling system, disconnect upper radiator hose and remove heat indicator sending unit from water outlet elbow. Remove injector lines and injectors. Remove intake and exhaust manifolds and the external oil feed line leading to cylinder head. Remove the rocker arm cover and rocker arms assembly, then unbolt and remove the cylinder head.

Mushroom type cam followers operate directly in machined bores in the cylinder head and are prevented from falling out as head is removed by the tappet adjusting screw lock nuts.

When installing the cylinder head, coat both sides of head gasket with a non-hardening sealing compound.

Gasket is marked "Top Front" for proper installation. Tighten cylinder head stud nuts to a torque of 55-60 ft.-lbs. using the sequence shown in Fig. 96 for three cylinder engine; or Fig. 97 for Model MF165. Adjust valve tappet gap as outlined in paragraph 104. Head should be re-torqued and valves adjusted after engine is warm.

VALVES AND SEATS
All Models

99. Intake and exhaust valves seat directly in the cylinder head. Valve heads and seat locations are numbered consecutively from front to rear. Any replacement valves should be so marked prior to installation. Intake and exhaust valves have a face angle

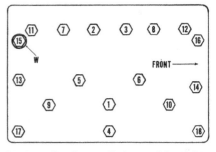

Fig. 96—On three cylinder models, tighten cylinder head stud nuts to a torque of 55-60 ft.-lbs. in the sequence shown. The washer (W) should be installed in the location shown.

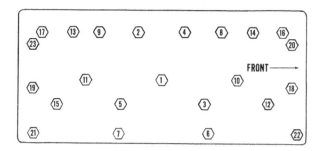

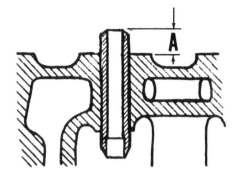

Fig. 97 — On Model MF165 diesel, tighten cylinder head stud nuts to a torque of 55-60 ft.-lbs. using the sequence shown.

Fig. 98—On all diesel models, valve guide is correctly positioned when distance (A) measures 0.584 - 0.594.

of 44 degrees, a seat angle of 45 degrees and a desired seat width of 1/16-inch. Seats can be narrowed using 20 and 70 degree stones.

Valve heads should be recessed a specified amount into the cylinder head. Clearance can be measured using a straight edge and feeler gage. Production clearances are held within the limits of 0.061-0.074. A maximum clearance of 0.140 is permissible before renewing valves or cylinder head, providing other conditions of satisfactory valve service are met. Inlet and exhaust valve tappet gap should be set to 0.010, hot.

VALVE GUIDES
All Models

100. The pre-sized intake and exhaust valve guides are interchangeable and can be pressed from the cylinder head if renewal is indicated. Install guide with counterbored end toward valve head and press guide into head until distance (A—Fig. 98) between top end of guide and spring seat on cylinder head measures 0.584-0.594.

Desired diametral clearance for the 0.311-0.312 valve stem in guide is 0.002-0.0045 for both intake and exhaust valves.

VALVE SPRINGS
All Models

101. Springs, retainers and locks are interchangeable for the intake and exhaust valves. Inner and outer springs are used as shown in Fig. 99. Springs may be installed with either end up. The inner spring (4) has a shorter assembled length than outer spring (5), due to the seating washer (3) and the machined step on spring retainer (6). Renew the springs if they are distorted, discolored, or fail to meet the test specifications which follow:

INNER SPRING:
Approx. free length....1⅜ inches
Lbs. test @ 1-3/16 inches......7-9
Lbs. test @ 27/32 inch 21-25

OUTER SPRING:
Approx. free length..1-25/32 inches
Lbs. test @ 1½ inches 21-25
Lbs. test @ 1-5/32 inches....48-52

CAM FOLLOWERS
All Models

102. The mushroom type tappets (cam followers) operate directly in machined bores in the cylinder head. The 0.6223-0.6238 diameter cam followers are furnished in standard size only and should have a diametral clearance of 0.0008-0.0035 in cylinder head bores.

To remove the cam followers after cylinder head is off, first remove the adjusting screw and locknut, then withdraw the cam follower from its bore.

ROCKER ARMS
All Models

103. The rocker arms and shaft assembly can be removed after removing the hood, fuel tank and rocker arm cover. The rocker arms are right hand and left hand units and should be installed on shaft as shown in Fig. 100 or 101. Desired diametral clearance between new rocker arms and new shaft is 0.0008-0.0035. Renew shaft and/or rocker arm if clearance is excessive.

The amount of oil circulating to the rocker arms is regulated by the rotational position of the rocker shaft in the support brackets. This position is indicated by a slot in one end of the rocker shaft as shown in Fig. 102. When the slot is positioned horizon-

tally, the maximum oil circulation is obtained. In production, the slot is positioned 30° from the vertical and the position indicated by a punch mark (P) on the adjacent support bracket. When reassembling, position the rocker shaft slot as indicated by the punch mark and check the assembly for proper lubrication. The shaft will not normally need to be moved from the marked position. The indicating slot is located on front end of shaft on three cylinder models, rear end of shaft on Model MF165.

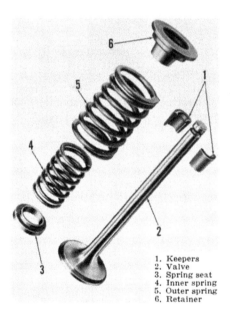

1. Keepers
2. Valve
3. Spring seat
4. Inner spring
5. Outer spring
6. Retainer

Fig. 99—Disassembled view of valve, showing correct positioning of springs and retainers.

Fig. 100 — Assembled view of rocker arm shaft for three cylinder engines.

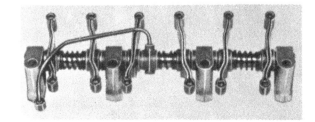

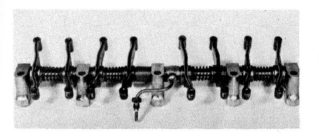

Fig. 101—Assembled view of rocker arm shaft for MF165 diesel.

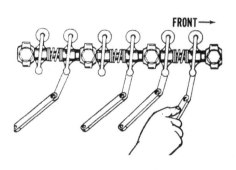

Fig. 102—End view of rocker shaft and support, showing indicating slot (S) and punch mark (P) correctly positioned for proper lubrication. Maximum lubrication is provided when slot (S) is horizontal.

VALVE TAPPET GAP
All Models

104. The recommended cold tappet gap setting is 0.012 for both the intake and exhaust valves of all models. Cold (static) setting of all valves can be made from just two crankshaft positions, using the procedure outlined in Figs. 103 and 104 for three cylinder engines; or Figs. 106 and 107 for Model MF165. Proceed as follows:

Remove timing plug from left side of flywheel adapter housing and turn crankshaft until the "TDC" timing mark is aligned with pointer as shown in Fig. 105.

On three cylinder models, check the No. 4 rocker arm counting from front of engine. If valve is open, No. 1 piston is on the compression stroke; adjust the four tappets shown in Fig. 103. If No. 4 valve is closed, No. 1 piston is on the exhaust stroke, adjust the two tappets shown in Fig. 104. Turn the crankshaft one complete turn until "TDC" timing mark is again aligned, then adjust the remaining tappets.

On Model MF165, check the rocker arms for the front and rear cylinders. If rear rocker arms are tight and

front rocker arms loose, No. 1 piston is on the compression stroke; adjust the four tappets shown in Fig. 106. If front rocker arms are tight and rear rocker arms loose, No. 4 piston is on the compression stroke; adjust the four tappets shown in Fig. 107. Turn the crankshaft one complete turn until "TDC" timing mark is again aligned, then adjust the remaining tappets.

On all models recheck the adjustment, if desired, with engine running at slow idle speed. Recommended tappet clearance with engine at operating temperature is 0.010. Clearance may be adjusted with engine running at slow idle speed or by following the procedure recommended for initial adjustment.

VALVE TIMING
All Models

105. To check the valve timing when engine is assembled, turn crankshaft until No. 1 piston is at TDC on the compression stroke; then, adjust clearance of front valve to 0.045. Insert a 0.002 feeler gage in front tappet gap and slowly turn engine in normal direction of rotation until feeler gage is tight. At this time, the "TDC" mark on flywheel should be aligned with timing pointer as shown in Fig. 105. NOTE: Timing may be considered correct if "TDC" mark is within ⅓-inch of alignment with timing pointer. Readjust the front valve to the recommended 0.012 before attempting to start the engine.

TIMING GEAR COVER
All Models

106. To remove the timing gear cover, drain the cooling system and remove the hood. On Model MF165, and Model MF150 with power steering, remove power steering pump from engine mounting and lay pump forward on front support. On all models, disconnect drag links, and radius rods if so equipped. Disconnect upper and lower radiator hoses and air cleaner hoses. Support the tractor under transmission case, unbolt front support from engine block and

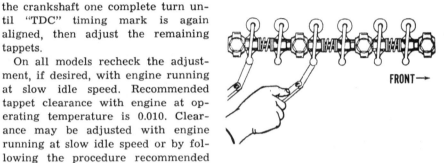

Fig. 103 — With "TDC-1" timing marks aligned as shown in Fig. 105 and No. 1 piston on compression stroke, adjust the indicated valves on three cylinder engines. Recommended cold tappet gap is 0.012. With four valves adjusted, turn crankshaft one revolution and refer to Fig. 104.

Fig. 104 — With "TDC-1" timing marks aligned as shown in Fig. 105 and No. 1 piston on exhaust stroke, adjust the indicated valves on three cylinder engines. Refer also to Fig. 103.

roll front support, front axle and radiator as an assembly away from tractor.

On model MF165, the front oil seal retainer (3—Fig. 109) can be removed after removing crankshaft pulley, without removing main timing gear cover. To renew the seal, remove the crankshaft pulley and loosen the two cap screws (C—Fig. 110), then, unbolt and remove the seal retainer as shown.

Fig. 105—"TDC-1" timing marks aligned as recommended for valve adjustment as outlined in paragraph 104.

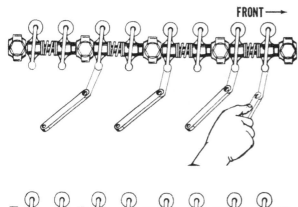

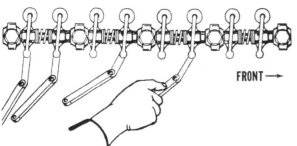

Fig. 106—On Model MF-165 diesel, with "TDC-1" timing marks aligned as shown in Fig. 105 and No. 1 piston on compression stroke, adjust the indicated valves to 0.012. With adjustment completed, turn crankshaft one revolution and refer to Fig. 107.

Fig. 107—On Model MF-165 diesel, with "TDC-1" timing marks aligned as shown in Fig. 105 and No. 4 piston on compression stroke, adjust the indicated valves to 0.012. With adjustment completed, turn crankshaft one revolution and refer to Fig. 106.

1. Upper cover
2. Inspection cover
3. Front seal retainer
4. Lower cover

Fig. 109—Front view of MF165 diesel engine showing assembled timing gear cover. Lower portion (4) is removed with oil pan.

Remove and discard the "O" ring (R—Fig. 111). Press the seal (S) into retainer from the front with seal lip to the rear, until front of seal is 3/32-inch from front face of bore. Install a new "O" ring (R), lubricate the "O" ring and use the crankshaft pulley (P—Fig. 112) as a pilot when reinstalling the seal retainer.

To remove the timing gear cover on all models, remove crankshaft pulley, fan belt and water pump.

The front oil seal (4—Fig. 113) used on three cylinder models can be renewed after timing gear cover (2) is off. Install seal from front of bore with seal lip to rear, until front face

of seal is ¼-inch beyond front face of timing gear cover. Camshaft thrust spring (1) is riveted to rear face of timing gear cover and controls camshaft end play.

TIMING GEARS

Three Cylinder Models

107. Fig. 114 shows a view of timing gear train with cover removed. Before attempting to remove any of the timing gears, first remove fuel tank, rocker arm cover and rocker arms assembly to avoid the possibility of damage to pistons or valve train if

Fig. 110—Lower cover, including the two clamp screws (C), should be loosened when installing oil seal retainer (3).

Fig. 111—Oil leakage around outside of retainer (3) is prevented by "O" ring seal (R). Lip seal (S) should be installed from front, with sealing lip to rear.

Fig. 108 — Adjusting the valve tappet gap on diesel engine.

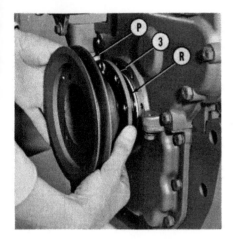

Fig. 112—Use the crankshaft pulley (P) as a pilot when installing front seal retainer (3). Lower timing gear cover should be loosened to prevent damage to "O" ring (R).

Fig. 113—Exploded view of timing gear cover and associated parts used on three cylinder models. Spring (1) controls camshaft end play.

1. Spring
2. Timing gear cover
3. Inspection cover
4. Oil seal

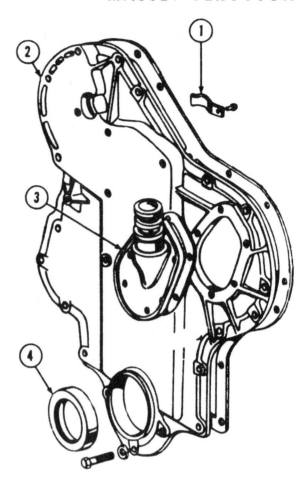

camshaft or crankshaft should be turned independently of the other.

Timing gear backlash should be 0.003-0.006 between the large idler gear and any of the other gears in the timing train. Replacement gears are available in standard size only. If backlash is not within specified limits, renew the idler gear, idler gear shaft and/or any other gears concerned.

To remove the timing gears or time the engine, unstake and remove the idler gear retaining bolt and slip the gear off idler shaft. The idler shaft is a light press fit in timing gear housing and is further positioned by the locating pin shown in Fig. 115. Pry the shaft from its place in timing gear housing if renewal is indicated.

The crankshaft gear is keyed in place and fits the shaft with a transition fit (0.001 tight to 0.001 loose). If the old gear is a loose fit, it may be possible to pry it off the shaft with a heavy screw driver or light pry bar. If a puller is needed, it will first be necessary to remove the oil pan and small lower section of timing gear housing.

The camshaft gear and injection pump drive gear can be removed by removing the cap screws and withdrawing the gears. To install the gears and time engine, refer to the appropriate following paragraphs:

108. **CAMSHAFT GEAR.** The gear is attached to camshaft by three equally spaced cap screws which thread into camshaft flange. It is possible, therefore, to install the gear in three positions, only one of which is correct. To correctly install the gear, align

the stamped letters "D" on camshaft hub and front face of gear as shown at (X—Fig. 114).

109. **INJECTION PUMP DRIVE GEAR.** The injection pump drive gear is retained to the pump adapter by three cap screws. When installing the gear, align dowel pin (2—Fig. 116) with slot (1) in adapter hub, then install the retaining cap screws. The injection pump drive gear and adapter are supported by the injection pump rotor bearings.

110. **TIMING THE GEARS.** Due to the odd size of the idler gear, the timing marks will align only once in eighteen crankshaft revolutions. To time the engine after the gears are removed, first install camshaft gear as outlined in paragraph 108 and the injection pump gear as in paragraph 109. Install the crankshaft gear on keyed shaft with timing mark to the front. Refer to Fig. 114 and turn crankshaft, camshaft and injection pump until the respective timing marks (T) point approximately toward idler gear hub; then, install idler gear with the retaining washer, locking washer and cap screw.

Model MF165

111. Fig. 117 shows a view of the timing gear train with cover removed and timing marks aligned. Due to the odd number of teeth in the two idler gears, all timing marks will align only once in more than 2,000 engine revolutions; therefore, timing is difficult to check without removing one or both idler gears.

Fig. 114—Three cylinder diesel engine timing gear train. Timing marks (T) must be aligned when engine is assembled, but will only align once in 18 crankshaft revolutions.

Fig. 115—Oil holes in cylinder block and idler gear stud are properly aligned by the locating pin shown.

Fig. 117 — Timing gear train for Model MF165 diesel engine. The timing marks (T & P) must be aligned when engine is assembled, but will align only once in more than 2000 crankshaft revolutions. Nonalignment of timing marks on idler gears (3 & 4) does not necessarily mean, therefore, that gears are incorrectly installed or engine is improperly timed.

1. Camshaft gear
2. Injection pump gear
3. Idler gear
4. Idler gear
5. Crakshaft gear
6. Idler gear
7. Oil pump drive gear

Before attempting to remove any of the timing gears, first remove fuel tank, rocker arm cover and rocker arms assembly to avoid the possibility of damage to pistons or valve train if camshaft or crankshaft should be turned with part of the gears removed.

Recommended timing gear backlash is 0.003-0.006 between any two gears in the timing gear train. Replacement gears are available in standard size only. If backlash is not within the specified limits, renew idler gears, idler gear shafts and/or the other gears concerned.

Refer to the appropriate following paragraphs for removal, installation and timing of the gears:

112. **IDLER GEARS AND HUBS.** Idler gears should have a diametral clearance of 0.003-0.0047 on idler gear hubs. Idler hub retainer plates and the flanged bushings in the lower idler gear (4—Fig. 117) are renewable. End play of assembled idler gear should be 0.002-0.004. Hubs (H—Fig. 119) are interchangeable for the two idler gears. Retaining stud holes permit hubs to be installed only one way, with oil holes (B) aligned. Idler hubs are a light press fit in engine block and can be pried out if renewal is indicated. Refer to paragraph 116 for installation and timing of idler gears.

113. **CAMSHAFT GEAR.** The cap screw holes in camshaft and camshaft gear (1—Fig. 117) are evenly spaced and the gear can be installed in three positions; only one of which is correct. The camshaft gear is correctly installed when the stamped "D" (Arrow—Fig. 120) on camshaft gear is aligned with the stamped "D" (Arrow—Fig. 121) on camshaft mounting flange.

114. **INJECTION PUMP DRIVE GEAR.** The injection pump drive gear (2—Fig. 117) is retained to the pump adapter by three cap screws. When installing the gear, align dowel pin

Fig. 118—Oil drilling (A) in idler gear (3) aligns with oil feed passage (B—Fig. 119) during gear rotation to provide lubrication for timing gear teeth.

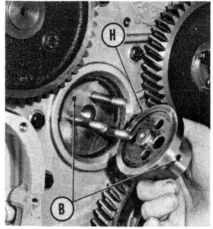

Fig. 119—Oil feed passage (B) in cylinder block and idler gear hub (H) must be aligned when hub is installed.

Fig. 116—Correct installation of injection pump drive gear is simplified by the dowel pin (2) which fits in machined notch (1) in pump drive shaft.

Fig. 120 — Retaining cap screw holes are evenly spaced in camshaft and camshaft timing gear. Gear is correctly installed when the stamped "D" mark (arrow) is aligned with similar mark on camshaft mounting flange. Refer to Fig. 121.

Fig. 121—The arrow indicates the stamped "D" timing mark on camshaft flange which must be aligned with similar mark on gear shown in Fig. 120.

Fig. 122—Dowel (D) in pump drive gear (G) fits in milled slot (S) in injection pump drive shaft (P) for correct pump timing. Timing marks (T) in the entire timing gear train must be aligned as shown in Fig. 117 when gear is installed.

(D—Fig. 122) with the slot (S) in pump adapter (P), then install the retaining cap screws. The injection pump drive gear and adapter are supported by the injection pump rotor bearings.

115. **CRANKSHAFT GEAR.** The crankshaft timing gear (5—Fig. 117) is keyed to the shaft and is a transition fit (0.001 tight to 0.001 loose) on shaft. It is usually possible to remove the gear using two small pry bars to move the gear forward. Use a suitable puller and adapter if gear cannot be removed with pry bars.

116. **TIMING THE GEARS.** To install and time the gears, first make sure camshaft gear, injection pump drive gear and crankshaft gear are correctly installed as outlined in paragraphs 113, 114 and 115. Turn crankshaft until key is pointing up as shown in Fig. 117 and install lower idler (4) with one timing mark aligned with mark on crankshaft gear and the other mark pointing toward upper idler gear hub. Turn camshaft gear (1) and pump drive gear (2) so timing marks point toward upper idler gear hub; then install upper idler gear (3) with all timing marks aligned. When timing marks are properly aligned, reinstall idler gear retainer washers, locks and stud nuts.

TIMING GEAR HOUSING

All Models

117. To remove the timing gear housing, first remove timing gears as outlined in paragraph 107 or 111. On three cylinder models, remove oil pan and the small section of housing extending below the crankshaft. On all

models remove the injection pump. Remove power steering pump if not previously removed. Block up the cam followers and withdraw camshaft; then, unbolt and remove timing gear housing. A small, cup-type expansion plug which closes front end of oil gallery is accessible after removing the housing.

Install by reversing the removal procedure.

CAMSHAFT

All Models

118. To remove the camshaft, first remove timing gear cover as outlined in paragraph 106, then remove fuel tank, rocker arm cover and rocker arms assembly. Secure the valve tappets (cam followers) in their uppermost position, remove the fuel lift pump, then withdraw camshaft and gear as a unit as shown in Fig. 123.

On three cylinder models, camshaft end float is dampened by the leaf-type thrust spring (1—Fig. 113) riveted to rear face of timing gear cover. Because of the spur-cut timing gears, end thrust does not present a problem. On Model MF165, camshaft end thrust of 0.003-0.006 is controlled by a machined boss on the timing gear cover and a renewable thrust washer (A—Fig. 124), located in front of engine block and positioned by locating pin (B).

The camshaft runs in three journal bores machined directly in engine block. The front and rear camshaft bearings are gravity lubricated by return oil from the rocker arms. The center journal is pressure lubricated by an external oil line shown in Fig. 125. The center journal, in turn, meters oil to the rocker shaft and cylin-

Fig. 123—Camshaft and timing gear may be withdrawn as a unit as shown, refer to text.

Fig. 124—Camshaft thrust washer (A) is positioned in engine block by locating pin (B).

der head through a second short oil feed line.

Camshaft bearing journals have a recommended diametral clearance of 0.004-0.008 in all three bearing bores. Journal diameters are as follows:

Front 1.869-1.870
Center 1.859-1.860
Rear 1.839-1.840

ROD AND PISTON UNITS

All Models

119. Connecting rod and piston units are removed from above after removing cylinder head, oil pan and rod bearing caps. Cylinder numbers are stamped on the connecting rod and cap. When reinstalling rod and piston units, make certain the correlation numbers are in register and face away from camshaft side of engine. Refer to Fig. 126. Tighten the connecting rod nuts to a torque of 45-50 ft.-lbs.

PISTONS, SLEEVES AND RINGS

All Models

120. The aluminum alloy, cam ground pistons are supplied in standard size only and are available in a kit consisting of piston, pin and rings for one cylinder. The toroidal combustion chamber is offset in piston crown and piston is marked (A—Fig. 126) for correct assembly.

Each piston is fitted with a plain faced chrome top ring which may be installed either side up. The plain faced cast iron, second compression ring may also be installed either side up.

The third compression ring consists of four steel segments as shown in Fig. 127. NOTE: Segments appear practically flat when not under compression; recommended installation

procedure is as follows: Grasp and compress the segment as shown in Fig. 128 until ring ends slightly overlap. When compressed, ring ends will curl up, as shown, or down if segment is turned over. Ring ends should curl down on segment placed in bottom of groove, up on second segment, down on third segment and up on top segment. Space end gaps 180° apart for alternate segments after all are installed.

The fourth groove uses a chrome plated, segmented oil ring, expander type. Stagger end gaps in the chrome rails.

The lower groove contains a cast iron oil ring which may be installed either side up.

Recommended end gap for all except segmented rings is 0.009-0.013; recommended side clearance is 0.002-0.004. Recommended piston skirt clearance, measured 90° from piston pin is 0.005-0.007.

The 0.0425 thick cast iron cylinder sleeves fit 0.001 loose to 0.001 tight in the 3.6875-3.6885 block bores. When installing the sleeves, make certain that the block bore (including the counterbore) and outside of cylinder sleeve is absolutely clean and dry. Chill the sleeve and press fully into place by hand. When properly installed, top of sleeve must be flush to 0.004 below top face of cylinder block. Sleeves must be pulled using a suitable puller.

PISTON PINS

All Models

121. The 1.24975-1.2500 diameter floating type piston pins are retained in piston bosses by snap rings and are available in standard size only. The renewable connecting rod bushing must be final sized after installation

to provide a diametral clearance of 0.0005-0.0017 for the pin. Be sure the pre-drilled oil hole in bushing is properly aligned with hole in top of connecting rod when installing new bushings. The piston pin should have a thumb press fit in piston after piston is heated to 160° F.

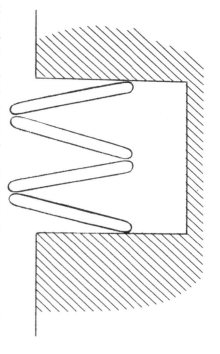

Fig. 127—Cross sectional view of segmented second compression ring correctly installed.

Fig. 125—External oil lines lead from main oil gallery to center camshaft journal, and from center journal to cylinder head for valve rocker lubrication. Valve lubrication is metered by the camshaft drillings.

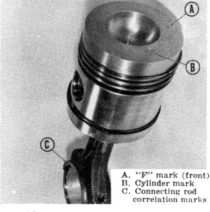

A. "F" mark (front)
B. Cylinder mark
C. Connecting rod correlation marks

Fig. 126—Assembled rod and piston unit showing location of marks.

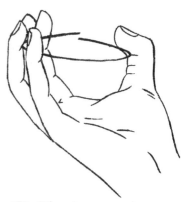

Fig. 128—When ring segment is compressed by hand as shown, ends of ring will point in direction which inner portion slants.

CONNECTING RODS AND BEARINGS

All Models

122. Connecting rod bearings are precision type, renewable from below after removing oil pan and bearing caps. When renewing bearing shells, be sure that the projection engages milled slot in rod and cap and that the correlation marks are in register and face away from camshaft side of engine. Replacement rods should be marked with the cylinder number in which they are installed. Bearings are available in standard, as well as undersizes of 0.010, 0.020 and 0.030.

Connecting rod bearings should have a diametral clearance of 0.0025-0.004 on the 2.2485-2.2490 diameter crankpin. Recommended connecting rod side clearance is 0.0095-0.015. Tighten the self-locking connecting rod nuts to a torque of 45-50 ft.-lbs.

CRANKSHAFT AND BEARINGS

All Models

123. The crankshaft is supported in precision type main bearings. Four main bearings are used on three cylinder models, five on Model MF165. To remove the rear main bearing cap on all models, it is first necessary to remove the engine, clutch, flywheel, flywheel adapter and rear oil seal. The remainder of the main bearing caps can be removed after removing the oil pan.

Upper and lower bearing liners are interchangeable on all except the front main bearing. When renewing the front bearing, make sure the correct half is installed in engine block. Bearing inserts are available in standard size and undersizes of 0.010, 0.020 and 0.030.

Main bearing caps are numbered from front to rear as shown at (B—Fig. 129). Block and cap bores are machined with caps in place and caps cannot be interchanged. Block and caps are stamped with a serial number (A) when the machining operation is performed. When installing caps, cap serial number must be on same side of block as the block serial number. Bearing caps are aligned to block by ring dowels as shown at (A—Fig. 130).

Some early engines are fitted with shim washers and locking tabs under main bearing cap screw heads. Both items have been eliminated in late production engines. Both the locking tabs and shim washers can be discarded when engine is serviced; however, if either item is retained, both must be used. Tighten the main bearing cap screws to a torque of 90-

Fig. 130—Renewable thrust washers (B & C) control crankshaft end play. Main bearing caps are positively located by ring dowels (A).

Fig. 131 — Crankshaft end play can be measured with a feeler gage as shown.

Fig. 129—Matched block and bearing caps bear a stamped serial number (A). Bearing caps are numbered front to rear as shown at (B).

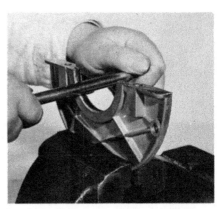

Fig. 132—Using a round bar to bed the asbestos rope seal in retainer half. Refer to text for details.

Fig. 133—On three cylinder models, crankshaft rear oil seal is housed in a two piece retainer bolted to cylinder block as shown. Cup plug (P) seals rear end of main oil gallery and oil leaks are possible if plug is not properly installed.

Fig. 134—Installed view of rear oil seal used on Model MF165 diesel. The four cap screws (C) are sealed with aluminum washers.

95 ft.-lbs. if shim washers and locking tabs are used; or 110-115 ft.-lbs. if washers and locks are discarded or not present.

Crankshaft end play is controlled by renewable thrust washers (B & C —Fig. 130) at front and rear of rear main bearing. The cap half of thrust washer is prevented from turning by the tab which fits in a machined notch in cap. Block half of washer can be rolled from position when bearing cap is removed. Recommended crankshaft end play is 0.002-0.014 for all models. The 0.123-0.125 thick thrust washers (B & C) are available in 0.007 oversize as well as standard.

When renewing rear main bearing, refer to paragraph 125 or 126 for rear oil seal installation.

Check the crankshaft journals against the values listed below:

Main journal diameter .. 2.7485-2.749
Crankpin diameter 2.2485-2.249
Diametral clearance
   Main bearings 0.003-0.005
   Crankpin bearings 0.0025-0.004

ENGINE ADAPTER PLATE

All Models

124. The engine flywheel is housed in a cast iron adapter plate which is located to the engine block by two dowels and secured to block rear face by six cap screws.

To obtain access to the rear main bearing, crankshaft or rear oil seal, it is first necessary to remove the adapter plate as follows:

Split tractor between engine and transmission and remove clutch and flywheel.

CAUTION: Flywheel is only lightly piloted to crankshaft flange and is not doweled.

Use care when removing to prevent flywheel from falling and causing possible injury.

After flywheel is off, remove the cap screws securing adapter plate to engine block and tap plate free of locating dowels. Install by reversing the removal procedure. Tighten flywheel retaining cap screws to a torque of 75 ft.-lbs.

REAR OIL SEAL

IMPORTANT: When rear oil leaks are present, check before and during disassembly, to make sure crankshaft seal is actually leaking. Other points of possible leakage are the cup plug (P—Fig. 133 or 136) located in rear face of engine block to seal the main oil gallery, seal retainer gaskets, or lower retainer cap screws (on MF165).

125. The asbestos rope type rear oil seal is contained in a two-piece seal retainer at rear of engine block. The seal retainer can be removed after removing flywheel and adapter plate.

The rope type crankshaft seal is precision cut to length, and must be installed in retainer halves with 0.010-0.020 of seal ends projecting from each end of retainer. To install the seal, clamp each half of retainer in a vise as shown in Fig. 132. Make sure seal groove is clean. Start each end of seal in groove with the specified amount of free end protruding. Allow seal rope to buckle in the center until about an inch of each end is bedded in groove, work center of seal into position, then roll with a round bar as shown. Repeat the process with the other half of seal. Install seal retainer as outlined in paragraph 126 or 127.

Three Cylinder Models

126. To install the oil seal, first make sure gasket surfaces of block and bearing cap are clean. Coat both sides of retainer gasket and end joints

Fig. 135—When installing rear oil seal housing on four cylinder models, use a straight edge as shown, to make sure rear edge is flush with block rear face.

of retainer halves with a suitable gasket cement. Coat shaft surface of rope seal with graphite grease. Install retainer halves and cap screws loosely; and tighten clamp screws thoroughly before tightening the retaining cap screws.

Model MF165

127. Use a straightedge as shown in Fig. 135 when installing oil pan seal housing, to make sure rear face of housing and block are flush. Coat both sides of retainer gasket and end joints of retainer halves with a suitable gasket cement. Coat shaft surface of rope seal with graphite grease. Install retainer halves and all cap screws loosely and tighten clamp screws thoroughly before tightening the retaining cap screws. The four lower cap screws (C—Fig. 134) are sealed with aluminum washers to prevent oil seepage at cap screw threads. Make sure aluminum washers are in good condition and properly installed.

Fig. 136—to renew rear oil seal (S) on four cylinder models, the oil pan, flywheel and engine adapter plate must be removed as shown. Check the cup plug (P) in oil gallery for signs of leakage while adapter plate is off.

Fig. 137—Checking flywheel runout with dial indicator. Refer to text for details.

Fig. 138 — Before removing oil pan on Model MF165 diesel, it is first necessary to remove front end unit and stud nut (N) from each side of timing gear housing.

FLYWHEEL

All Models

128. To remove the flywheel, first separate the engine from transmission housing and remove the clutch. Flywheel is secured to crankshaft flange by six evenly spaced cap screws. To properly time flywheel to engine during installation, be sure that unused hole in flywheel aligns with untapped hole in crankshaft flange.

CAUTION: Flywheel is only lightly piloted to crankshaft. Use caution when unbolting flywheel to prevent flywheel from falling and causing possible injury.

The starter ring gear can be renewed after flywheel is removed. Heat ring gear evenly to 475°-500° F. and install on flywheel with beveled end of teeth facing front of engine.

Check flywheel runout with a dial indicator as shown in Fig. 137 after flywheel is installed. Maximum allowable flywheel runout must not exceed 0.001 for each inch from flywheel centerline to point of measurement. Tighten the flywheel retaining cap screws to a torque of 75 ft.-lbs. when installing the flywheel.

OIL PAN

All Models

129. The heavy cast-iron oil pan serves as the tractor frame and attaching point for the tractor front support. To remove the oil pan, first drain cooling system and oil pan. Remove hood and side panels. On MF135, disconnect drag links and radius rods. On other models disconnect drag link

and if so equipped, unbolt power steering pump from engine and lay pump forward in front support. On all models, support tractor underneath transmission housing, disconnect radiator hoses and air cleaner hose, unbolt front support from oil pan and roll front support, front axle and radiator as an assembly away from tractor.

Place a rolling floor jack under oil pan. On three cylinder models, remove the attaching cap screws and lower the oil pan away from engine block. On Model MF165, refer to Fig. 138 and remove stud nut (N) at left front corner of oil pan. Remove the similar stud nut on right side of block which is located behind front support spacer (S—Fig. 139). Remove the two cap screws (C) from timing gear cover, remove remainder of oil pan retaining cap screws and lower the pan with lower timing gear cover (L) attached.

Before reinstalling the oil pan on Model MF165, remove lower timing gear cover (L), and reinstall finger tight using a new gasket. Examine the "O" ring on front oil seal retainer and, if in good condition, reinstall pan and tighten the cap screws retaining pan to engine block and flywheel adapter. Install and tighten the two cap screws (C); then tighten the screws retaining lower timing gear housing (L) to oil pan.

OIL PUMP

All Models

130. The rotary type oil pump is mounted on front main bearing cap

Fig. 139—The front support spacer (S) must be removed for access to hidden stud nut. Remove cap screw (C) from each side of timing gear cover and allow lower cover (L) to remain on oil pan.

Fig. 140—Mounted view of engine oil pump and drive assembly with oil pan removed. Four cylinder engine is shown but three cylinder unit is similar.

Fig. 141—Removing engine oil pump as an assembly with front main bearing cap.

Fig. 144—Rotor to body clearance should not exceed 0.010.

Fig. 145 — Rotor end clearance can be measured with a straight-edge and feeler gage as shown. Clearance should not exceed 0.003.

Fig. 142—Engine oil pump may be unbolted from front main bearing cap after removing idler gear.

and driven from the crankshaft timing gear through an idler as shown in Fig. 140.

To remove the oil pump, first remove oil pan as outlined in paragraph 129. On three cylinder models, remove the small lower section of timing gear housing which extends below crankshaft and seals front of oil pan. Pump can be removed as an assembly with front bearing cap as shown in Fig. 141, or detached from cap as shown in Fig. 142 after removing idler gear.

Check rotor clearance with a feeler gage as shown in Fig. 143. Clearance should not exceed 0.006. Rotor to body clearance should not exceed 0.010 when checked as shown in Fig. 144. Check rotor end clearance with a straight edge and feeler gage as shown in Fig. 145; clearance should not exceed 0.003. Except for the drive idler gear, relief valve plunger and relief valve spring, the oil pump is available only as an assembly. If rotor or body are scored, otherwise damaged, or fail to meet the test specifications listed above, renew the pump.

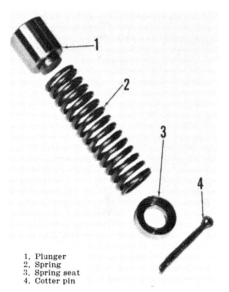

1. Plunger
2. Spring
3. Spring seat
4. Cotter pin

Fig. 146 — Disassembled view of plunger type relief valve which is located in oil pump body.

RELIEF VALVE

All Models

131. The plunger type relief valve (Fig. 146) is located in oil pump body, and can be adjusted by renewing spring (2) to maintain a relief pressure of 50-65 psi at operating speed. Spring (2) should have a free length of 1½-inches and should test 5¼-7¼ lbs. at 1¼-inches or 11½-13½ lbs. at 1 inch.

Fig. 143—Rotor gear clearance should not exceed 0.006 when measured with a feeler gage as shown.

GASOLINE FUEL SYSTEM

CARBURETOR

All Models

132. ADJUSTMENT. Marvel-Schebler or Zenith carburetors may be used. Initial adjustments are approximately one turn open for idle mixture adjustment needle and 1¼-1½ turns open for main adjustment needle. Final adjustments must be made under operating conditions with engine at normal temperature. After mixture adjustments have been made, adjust low idle speed to 450-500 rpm for models with Continental engine or 725-775 rpm for models with Perkins engine.

133. OVERHAUL (MARVEL-SCHEBLER). Refer to Fig. 147. To disassemble the removed carburetor, first clean outside with a suitable solvent and remove main adjusting needle (14) and idle mixture adjusting needle (15). Remove the four screws which retain throttle body (13) to fuel bowl (3) and lift off

throttle body, gasket and venturi (7). Remove float shaft and float, gasket, venturi and inlet needle valve. Withdraw venturi from gasket and discard the gasket. Remove inlet needle valve seat, idle jet (12) and main nozzle (6). Discard the gaskets from nozzle and needle valve seat.

Remove throttle and choke valves, shafts and packing. Bushings for throttle and choke shafts are not provided.

Clean all parts in a suitable carburetor cleaner and rinse in clean mineral solvent. Blow out passages in body and bowl with compressed air and renew any parts which are worn or damaged.

Assemble by reversing the disassembly procedure, using new gaskets and packing. Install throttle valve (11) with marking toward mounting flange on the side away from adjusting needles. Adjust float height to ¼-inch when measured from gasket surface to nearest edge of float, with throttle body inverted. If adjustment is required, carefully bend float arms using a bending tool or needle nose pliers, keeping the two halves of float parallel and equal. Check mixture adjustments after installation, as outlined in paragraph 132.

134. OVERHAUL (ZENITH). To disassemble the removed carburetor, first clean outside with a suitable solvent. Remove the screws retaining throttle body (16—Fig. 148) to fuel bowl (8) and remove fuel bowl. Remove float shaft, float (12) and inlet valve needle. NOTE: Float shaft is a tight fit in slotted side of hinge bracket and should be removed from opposite side.

Remove venturi (11), inlet valve seat and idle jet (14) from throttle body. Remove idle adjusting needle (18). Remove main adjusting needle assembly (1), discharge jet (2), main jet (9) and well vent jet (10) from fuel bowl. Remove throttle and choke valves, shafts and packing. Bushings for throttle and choke shafts are not provided.

Discard all gaskets and packing and clean remainder of parts in a suitable carburetor cleaner. Rinse in clean mineral solvent and blow out passages in body and bowl with compressed air. Renew all gaskets and packing and any other parts which are worn or damaged.

Assemble by reversing the disassembly procedure, using new gaskets and packing. Install throttle valve (17) so beveled edges will fit throttle body bore with throttle closed, with side of throttle plate farthest from mounting flange aligned with idle port. Adjust float height to $1\frac{5}{32}$ inch when measured from gasket surface of throttle body to farthest edge of float at free ends. If adjustment is required, carefully bend float arms using a bending tool or needle nose pliers, keeping the two halves of float parallel and equal. Check mixture adjustments after installation, as outlined in paragraph 132.

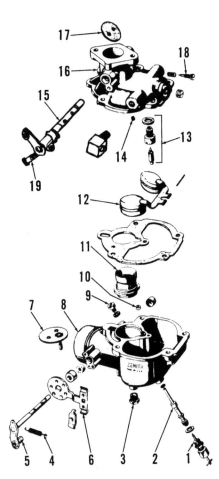

Fig. 148—Exploded view of Zenith carburetor used on some models.

1. Main adjustment needle	10. Well vent jet
2. Discharge jet	11. Venturi
3. Drain plug	12. Float
4. Spring	13. Inlet needle valve
5. Choke shaft	14. Idle jet
6. Bracket	15. Throttle shaft
7. Choke valve	16. Throttle body
8. Float chamber	17. Throttle valve
9. Main jet	18. Idle mixture needle
	19. Idle speed screw

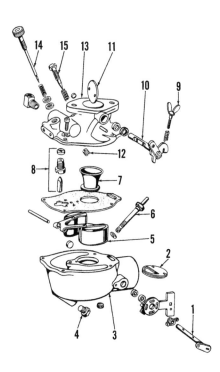

Fig. 147—Exploded view of Marvel-Schebler carburetor used on some models.

1. Choke shaft	9. Idle speed screw
2. Choke valve	10. Throttle shaft
3. Fuel bowl	11. Throttle valve
4. Drain plug	12. Idle jet
5. Float	13. Throttle body
6. Main nozzle	14. Main adjustment needle
7. Venturi	
8. Inlet needle valve	15. Idle mixture needle

DIESEL FUEL SYSTEM

The diesel fuel system consists of three basic units; the fuel tank and filters, injection pump and injector nozzles. When servicing any unit associated with the diesel fuel system, the maintenance of absolute cleanliness is of utmost importance. Of equal importance is the avoidance of nicks or burrs on any of the working parts.

Probably the most important precaution that service personnel can impart to owners of diesel powered tractors is to urge them to use an approved fuel that is absolutely clean and free from foreign materials. Extra precaution should be taken to make certain that no water enters the fuel sorage tanks. Because of the high pressures and degree of control required of injection equipment, extremely high precision standards are necessary in the manufacture and servicing of diesel components. Extra care in daily maintenance will pay big dividends in long service life and the avoidance of costly repairs.

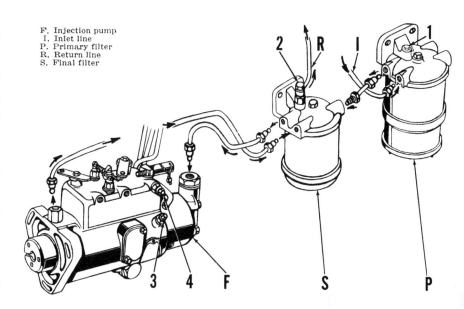

F. Injection pump
I. Inlet line
P. Primary filter
R. Return line
S. Final filter

Fig. 149—Schematic view of diesel injection pump, fuel filters and lines. Arrows indicate direction of fuel flow. Bleed screws and proper bleeding order are indicated by the numerical references (1, 2, 3 & 4).

FUEL FILTERS AND LINES

All Models

135. **OPERATION AND MAINTENANCE.** Refer to Fig. 149 for a schematic view of fuel flow through filters and injection pump.

NOTE: Actual location of filters may differ somewhat from that shown. The camshaft actuated, diaphragm type fuel lift pump is not shown.

A much greater volume of fuel is circulated within the system than is burned in the engine, the excess serving as a coolant and lubricant for the injection pump. Fuel enters the primary filter (P) through inlet line (I), where it passes through the water trap (Agglomerator) and first stage filter element. Both lines leading to injection pump (F), and return line (R), are connected to a common passage in secondary filter (S); and separated from primary filter line only by the secondary filter element. The greater volume of filtered fuel is thus recirculated between the secondary filter (S) and injection pump (F). A much smaller quantity of fuel enters the system through inlet line (I) or returns to the tank through line (R), thus contributing to longer filter life.

Inspect the glass bowl at bottom of primary filter (P) daily and drain off any water or dirt accumulation. Drain the primary filter at 100 hour intervals and renew the element each 500 hours. Renew element in secondary filter (S) every 1000 hours. Renew both elements and clean the tank and lines if evidence of substantial water contamination exists.

DIESEL SYSTEM TROUBLE-SHOOTING CHART

	Sudden Stopping of Engine	Lack of Power	Engine Hard to Start	Irregular Engine Operation	Engine Knocks	Excessive Fuel Consumption
Lack of fuel	★	★	★	★		
Water or dirt in fuel	★	★	★	★		
Clogged fuel lines	★	★	★	★		
Inferior fuel		★	★	★	★	★
Faulty primary pump	★	★	★	★		
Faulty injection pump timing		★	★	★	★	★
Air traps in system	★	★	★	★		
Clogged fuel filters	★	★	★	★		
Faulty nozzle		★		★	★	★
Faulty injection pump	★	★	★	★	★	★

Fig. 150—Fuel lift pump showing manual lever (M).

Fig. 151—A suitable injector tester is required to completely test and adjust the injector nozzles.

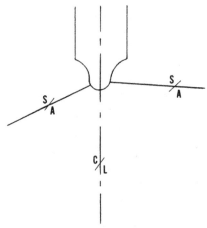

Fig. 153—Nozzle spray pattern is not symmetrical with centerline of nozzle tip.

Fig. 152—Nozzle holes (arrows) are not located an equal distance from nozzle tip.

136. **BLEEDING.** To bleed the system, make sure tank shut-off valve is open, have an assistant actuate the manual lever (M—Fig. 150) on fuel lift pump, and proceed as follows:

Loosen the air vent (1—Fig. 149) on primary filter (P) and continue to operate the lift pump until air-free fuel flows from vent plug hole. Tighten plug (1). Loosen vent plugs (2, 3 and 4) on secondary filter and injection pump, in turn while continuing to operate the lift pump. Tighten each plug as air is expelled and proceed to the next. Operate manual lever for approximately ten extra strokes after tightening vent plug (4), to expel any air remaining in bleed back lines.

With the fuel supply system airbled, push in the stop button, partially open throttle lever and attempt to start the tractor. If tractor fails to fire, loosen compression nut at all injector nozzles and turn engine over with starter until fuel escapes from all loosened connections. Tighten compression nuts and start engine.

TROUBLE SHOOTING

All Models

137. **QUICK CHECKS—UNITS ON TRACTOR.** If the diesel engine fails to start or does not run properly, and the diesel fuel system is suspected as the source of trouble, refer to the Diesel System Trouble Shooting Chart and locate points which require further checking. Many of the chart items are self-explanatory; however, if the difficulty points to the fuel filters, injection nozzles and/or injection pump, refer to the appropriate paragraphs covering that particular unit.

INJECTOR NOZZLES

All Models

WARNING: Fuel leaves the injector nozzle with sufficient pressure to penetrate the

Fig. 154—Exploded view of C. A. V. injector nozzle and holder assembly. Correct opening pressure is indicated on tab (4).

1. Cap nut	8. Valve spindle
2. Gasket	9. Nozzle holder
3. Locknut	10. Nozzle valve
4. Tab	11. Nozzle body
5. Adjusting screw	12. Nozzle nut
6. Adjusting shim	13. Seat washer
7. Spring	14. Dust shield

skin. Keep unprotected parts of your body clear of the nozzle spray when testing.

All models are equipped with C. A. V. multi-hole nozzles which extend through the cylinder head to inject fuel charge into a combustion chamber machined in crown of piston.

138. **TESTING AND LOCATING A FAULTY NOZZLE.** If rough or uneven engine operation, or misfiring, indicates a faulty injector, the defective unit can usually be located as follows:

With engine running at the speed where malfunction is most noticeable (usually low idle speed), loosen the compression nut on high pressure line for each injector in turn and listen for a change in engine performance. As in checking spark plugs, the faulty unit is the one which, when its line is loosened, least affects the running of the engine.

If a faulty nozzle is found and considerable time has elapsed since the injectors have been serviced, it is recommended that all nozzles be removed and serviced or that new or reconditioned units be installed. Refer to the following paragraphs for removal and test procedure.

139. **REMOVE AND REINSTALL.** Before loosening any fuel lines, thoroughly clean the lines, connections, injectors and engine area surrounding the injector, with air pressure and solvent spray. Disconnect and remove the leak-off line, disconnect pressure line and cap all connections as they are loosened, to prevent dirt entry into the system. Remove the two stud nuts and withdraw injector unit from the cylinder head.

Thoroughly clean the nozzle recess in cylinder head before reinstalling injector unit. It is important that seating surface be free of even the smallest particle of carbon or dirt which could cause the injector unit to be cocked and result in blow-by. No hard or sharp tools should be used in cleaning. Do not re-use the copper sealing washer located between injector nozzle and cylinder head, always install a new washer. Each injector should slide freely into place in cylinder head without binding. Make sure that dust seal is reinstalled and tighten the retaining stud nuts evenly to a torque of 20-22 ft.-lbs. After engine is started, examine injectors for blow-by, making the necessary correction before releasing tractor for service.

140. **TESTING.** A complete job of testing and adjusting the injector requires the use of special test equipment. Only clean, approved testing oil should be used in the tester tank. The nozzle should be tested for opening pressure, seat leakage, back leakage, and spray pattern. When tested, the nozzle should open with a sharp popping or buzzing sound and cut off quickly at end of injection with a minimum of seat leakage and controlled amount of back leakage.

Before conducting the test, operate tester lever until fuel flows, then attach the injector. Close the valve to tester gage and pump tester lever a few quick strokes to be sure nozzle valve is not plugged, that four sprays emerge from nozzle tip, and that possibilities are good that injector can be returned to service without overhaul.

NOTE: Spray pattern is not symmetrical with centerline of nozzle tip. The apparently irregular location of nozzle holes (See Figs. 152 and 153) is designed to provide the correct spray pattern in the combustion chamber.

If adjustment is indicated by the preliminary tests, proceed as follows:

141. OPENING PRESSURE. Open the valve to tester gage and operate tester lever slowly while observing gage reading. Opening pressure should be 2500 psi for three cylinder models or 2575 psi for Model MF165. If opening pressure is not as specified, remove the injector cap nut (1—Fig. 154), loosen locknut (3) and turn adjusting sleeve (5) as required to obtain the recommended pressure.

NOTE: When adjusting a new injector or overhauled injector with new pressure spring (7), set the pressure 220 psi higher than specified, to allow for initial pressure loss.

142. SEAT LEAKAGE. The nozzle tip should not leak at a pressure less than 2300 psi. To check for leakage, actuate tester lever slowly and as the gage needle approaches 2300 psi, observe the nozzle tip. Hold the pressure at 2300 psi for ten seconds; if drops appear or if nozzle tip is wet, the valve is not seating and the injector must be disassembled and overhauled as outlined in paragraph 145.

143. BACK LEAKAGE. If nozzle seat as tested in paragraph 142 is satisfactory, check the injector and connections for wetness which would indicate leakage. If no visible external leaks are noted, bring gage pressure to 2250 psi, release the lever and observe the time required for gage pressure to drop from 2250 psi to 1500 psi. For a nozzle in good condition, this time should not be less than six seconds. A faster pressure drop would indicate a worn or scored nozzle valve piston or body, and the nozzle assembly should be renewed.

NOTE: Leakage of the tester check valve or connections will cause a false reading, showing up in this test as excessively fast leak-back. If all injectors tested fail to pass this test, the tester rather than the units should be suspected as faulty.

144. SPRAY PATTERN: If leakage and pressure are as specified when tested as outlined in paragraphs 141 through 143, operate the tester handle several times while observing spray pattern. Four finely atomized, equally spaced, conical sprays should emerge from nozzle tip, with equal penetration into the surrounding atmosphere.

If pattern is uneven, ragged or not finely atomized, overhaul the nozzle as outlined in paragraph 145. NOTE: Spray pattern is not symmetrical with centerline of nozzle tip as shown in Fig. 153.

145. **OVERHAUL.** Hard or sharp tools, emery cloth, grinding compound, or other than approved solvents or lapping compounds must never be used. An approved nozzle cleaning kit is available through any C. A. V. Service Agency and other sources.

Wipe all dirt and loose carbon from exterior of nozzle and holder assembly. Refer to Fig. 154 and proceed as follows. Secure the nozzle in a soft-jawed vise or holding fixture and remove the cap nut (1). Loosen jam nut (3) and back off the adjusting sleeve (5) to completely unload the pressure spring (7). Remove the nozzle cap nut (12) and nozzle body (11). Nozzle valve (10) and body (11) are matched assemblies and must never be intermixed. Place all parts in clean calibrating oil or diesel fuel as they are disassembled. Clean the injector assembly exterior with a soft wire brush, soaking in an approved carbon solvent, if necessary, to loosen hard carbon deposits on exterior and in spray holes. Rinse the parts in clean diesel fuel or calibrating oil immediately after cleaning, to neutralize the carbon solvent and prevent etching of the polished surfaces. Clean the pressure chamber of nozzle tip using the special reamer as shown in Fig. 155. Clean the spray holes in nozzle with an 0.009 (0.024 mm) wire probe held in a pin vise as shown in Fig. 156. Wire probe should protrude from pin vise only far enough to pass through spray holes (approximately 1/16-inch), to prevent bending and breakage. Rotate pin vise without applying undue pressure.

Clean valve seats by inserting small end of brass, valve seat scraper into

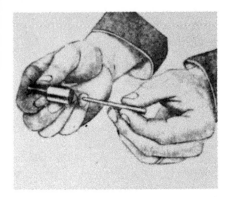

Fig. 155 — Clean the pressure chamber in nozzle tip using the special reamer as shown.

Fig. 156 — Clean the spray holes in nozzle tip using a pin vise and 0.009 wire probe.

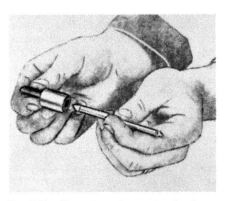

Fig. 157—Clean the valve seats using brass scraper as shown.

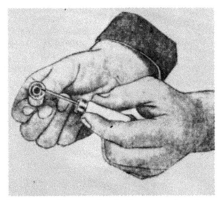

Fig. 158—Use the hooked scraper to clean annular groove in top of nozzle body.

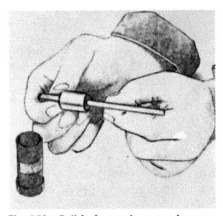

Fig. 159—Polish the nozzle seat using mutton tallow on a wood polishing stick.

Fig. 160—Make sure locating dowels are carefully aligned when nozzle body is reinstalled.

nozzle and rotating tool. Reverse the tool and clean upper chamfer using large end. Refer to Fig. 157. Use the hooked scraper to clean annular groove in top of nozzle body (Fig. 158). Use the same hooked tool to clean the internal fuel gallery.

With the above cleaning accomplished, back flush the nozzle by installing the reverse flusher adapter on injector tester and nozzle body in adapter, tip end first. Secure with the knurled adapter nut and insert and rotate the nozzle valve while flushing. After nozzle is back flushed, seat can be polished by using a small amount of tallow on end of polishing stick, rotating the stick as shown in Fig. 159.

Light scratches on valve piston and bore can be polished out by careful use of special injector lapping compound only, DO NOT use valve grinding compound or regular commercial polishing agents. DO NOT attempt to reseat a leaking valve using polishing compound. Clean thoroughly and back flush if lapping compound is used.

Reclean all parts by rinsing thoroughly in clean diesel fuel or calibrating oil and assemble valve to body while immersed in the cleaning fluid. Reassemble the injector while still wet. With adjusting sleeve (5—Fig. 154) loose, reinstall nozzle body (11) to holder (9), making sure valve (10) is installed and locating dowels aligned as shown in Fig. 160. Tighten nozzle cap nut (12—Fig. 154) to a torque of 50 ft.-lbs. Do not overtighten, distortion may cause valve to stick and overtightening cannot stop a leak caused by scratches or dirt on the lapped mating surfaces of valve body and nozzle holder.

Retest and adjust the assembled injector assembly as outlined in paragraphs 140 through 144.

NOTE: If overhauled injector units are to be stored, it is recommended that a calibrating or preservative oil, rather than diesel fuel be used for the pre-storage testing. Storage of more than thirty days may result in the necessity of recleaning prior to use of the unit.

INJECTION PUMP

All Models

The injection pump is a completely sealed unit. No service work of any kind can be accomplished on the pump or governor unit without the use of special, costly, pump testing equipment and special training. The only adjustment authorized is adjustment of the idle speed screw (I—Fig. 161). If additional service work is required, the pump should be turned over to an authorized C. A. V. service station for overhaul. Inexperienced or unequipped service personnel should never attempt to overhaul a diesel injection pump.

146. **ADJUSTMENT.** The slow idle stop screw (I—Fig. 161) should be adjusted with engine warm and running, to provide the recommended slow idle speed of 750 rpm for Model MF165; or 700 rpm for other models. Also check to make sure that governor arm contacts the slow idle screw (I) and high speed screw (H) when throttle lever is moved to slow and fast positions. Also check to make sure that stop lever arm (L—Fig. 162) moves fully to operating position when stop button is pushed in, and shuts off fuel to injectors when stop button is pulled. The high speed stop screw (H—Fig. 161) is set at the factory and the adjustment is sealed. Governed speed under load should be 2000 rpm for all models, with high-idle (no load) speed of approximately 2160 rpm.

Fig. 161—Injection pump showing linkage and speed adjusting screws.

H. Maximum speed
stop screw
I. Idle speed
stop screw
N. Nut
S. Stud
T. Throttle link

Fig. 164 — Locating slot (S) must align with dowel (arrow—Fig. 163) when pump is reinstalled. Pump timing cover is shown at (C).

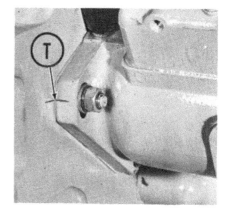

Fig. 165 — Timing marks (T) should be aligned when pump is reinstalled.

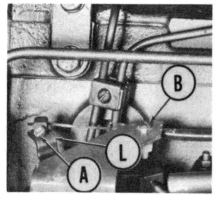

Fig. 162 — Injection pump showing stop cable and associated parts.

A. Cable screw
B. Clamp screw
L. Stop lever

Fig. 163—To remove the injection pump, it is first necessary to remove inspection cover and the three gear retaining cap screws as shown. Dowel (arrow) times the pump when unit is reinstalled.

Refer to paragraph 148 for pump timing adjustment.

147. **REMOVE AND REINSTALL.** Before attempting to remove the injection pump, thoroughly wash the pump and connections with clean diesel fuel or an approved solvent. Disconnect throttle control rod (T—Fig. 161) from governor arm by removing nut (N) and withdrawing stud (S) from arm. Disconnect stop control rod or cable from stop lever (L—Fig. 162), and on models so equipped, disconnect cable housing by loosening clamp screw (B). Remove the inspection cover from front of timing gear cover, then remove the three screws securing injection pump drive gear to pump as shown in Fig. 163. Disconnect fuel inlet, outlet and high pressure lines from pump, capping all connections to prevent dirt entry. Check to see that timing marks (T—Fig. 165) align, remove the three flange stud nuts, then withdraw the pump as shown in Fig. 164.

To install the pump, align the milled slot (S) with dowel (Arrow—Fig. 163) and insert the pump. Align timing marks (T—Fig. 165), then complete the installation by reversing the removal procedure. Bleed the system as outlined in paragraph 136. Check the injection pump timing, if necessary, as outlined in paragraph 148.

NOTE: Pump can only be installed in one position. If timing gear train has not been disturbed and timing marks (T—Fig. 165) are aligned, timing should be correct.

148. **PUMP TIMING TO ENGINE.** The injection pump drive shaft is fitted with a milled slot (S—Fig. 164) in forward end which engages in dowel pin (Arrow—Fig. 163) in pump drive gear. Thus, injection pump can

Fig. 166—On three cylinder engines, the "E" timing mark on injection pump rotor should align with scribe mark on snap ring when flywheel timing mark is aligned as shown in Fig. 167. Four cylinder models use "B" mark.

Fig. 167—Flywheel timing cover removed showing 24° BTDC timing mark aligned.

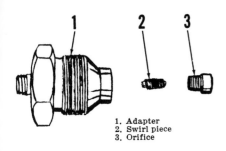

1. Adapter
2. Swirl piece
3. Orifice

Fig. 168—Exploded view of cold starting. unit manifold adapter.

(Red) to threads of orifice (3), re-install orifice and tighten to a torque of 25-30 ft.-lbs.

NOTE: If orifice (3) cannot be removed, heat adapter (1) in an oven or by other suitable means, to a temperature of approximately 500° F., and remove while hot. Heat is necessary to break the previously applied Loctite seal. Loctite is used to prevent orifice from loosening in adapter and entering engine through manifold.

NON-DIESEL GOVERNOR

be removed and reinstalled without regard to timing position. NOTE: Injection pump drive gear cannot become unmeshed from idler gear when timing gear cover is in place, therefore timing is not disturbed by removal and installation of pump.

To check the pump timing, shut off the fuel, remove timing window (C—Fig. 164) and flywheel timing plug from left side of flywheel adapter plate. Turn the crankshaft until 24° BTDC static timing mark is aligned in flywheel timing window as shown in Fig. 167; at which time correct timing mark on injection pump rotor should align with scribed line on lower end of snap ring as shown in Fig. 166. The mounting holes in pump flange are elongated to permit minor timing variations. If timing marks cannot be properly aligned by shifting pump on mounting studs, the timing cover must be removed as outlined in paragraph 106 and the gears retimed.

COLD STARTING UNIT

Accessory - All Models

149. Some tractors are optionally equipped with an ether injection type cold weather starting aid consisting of a dash-mounted fuel can adapter, connecting tube, and a manifold adapter (Fig. 168). The only service required is renewal of damaged parts or occasional cleaning if the unit should become inoperative. Proceed as follows:

Disconnect the connecting tubing from manifold adapter housing (1). Blow compressed air through the can adapter and tube to make sure they are open and clean. Remove adapter housing (1) from intake manifold, unscrew orifice plug (3) and withdraw swirl piece (2). Clean all parts in a suitable solvent and drop swirl piece (2) into bore of adapter (1). Apply a small amount of Loctite, Grade "A"

ADJUSTMENT

All Models

150. Recommended governed speeds are as follows:

Low Idle
  Continental Engine . . 450 - 500 rpm
  Perkins Engine . . . . . . 725 - 775 rpm
High Idle
  Continental Engine 2200 - 2250 rpm
  Perkins Engine . . . . 2225 - 2275 rpm
Loaded Speed
  All Models . . . . 2200 rpm

All models are equipped with an external spring connecting throttle control rod to governor arm, and adjustable linkage between governor arm and throttle. To adjust the governor, proceed as follows:

With engine not running, disconnect governor rod and carburetor link from governor arm. Move governor arm toward wide-open position (forward on MF135 & MF150 with Continental Engine, rearward on other

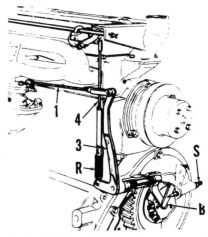

Fig. 169—Schematic view of governor and associated parts used on Model MF135 with Continental engine.

1. Carburetor link
3. Governor arm
4. Governor rod
B. Bumper spring
R. Governor spring
S. Bumper screw

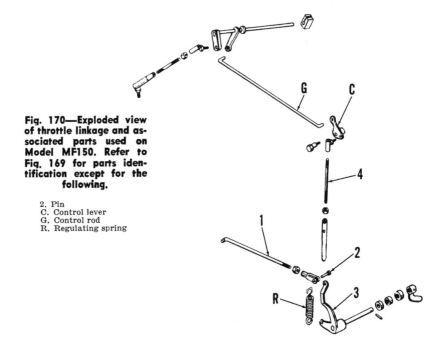

Fig. 170—Exploded view of throttle linkage and associated parts used on Model MF150. Refer to Fig. 169 for parts identification except for the following.

2. Pin
C. Control lever
G. Control rod
R. Regulating spring

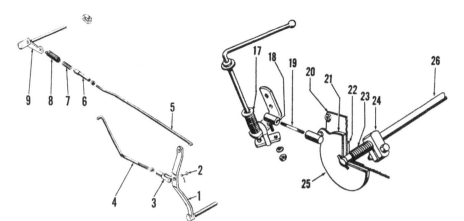

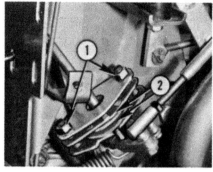

Fig. 171—Exploded view of throttle linkage and associated parts used on Model MF165 with Continental engine.

1. Governor arm	6. Plunger
2. Pin	7. Inner spring
3. Yoke	8. Spring
4. Carburetor rod	9. Throttle crank
5. Governor rod	

Fig. 172—Hand throttle lever and associated parts used on Model MF135 with Continental engine. To increase pressure on friction disc, loosen clamp (24) and move rearward on rod (26).

17. Spring	22. Washer
18. Ball joint	23. Spring
19. Control link	24. Clamp
20. Friction plate	25. Throttle Plate
21. Friction disc	26. Throttle rod

Fig. 173—On Models MF150 and MF165, throttle hand lever friction is adjusted by tightening nuts (1).

Fig. 175 — Camshaft mounted governor weight unit and associated parts of the type used on Model MF135 and Model MF150 with Continental engine. Corresponding parts on models with Perkins engine is similar except that governor is carried on distributor drive shaft gear.

1. Race and shaft
2. Camshaft nut
3. Weight unit
L. Lip

models). Move throttle link to wide-open position and adjust the clevis if necessary, until carburetor throttle reaches wide-open position just before governor arm completes its travel.

Reconnect governor rod and carburetor link. On Models MF135 and MF150 with Continental engine, loosen the locknut and back out bumper screw (S—Fig. 169) three or four turns to be sure it is free from spring (B).

On all models, move hand throttle lever to fast and slow positions and check to be sure governor arm moves through its full travel. Start and warm the engine. Adjust low engine speed at low idle speed screw on carburetor and high idle speed by shortening governor rod to apply more tension to governor spring.

If governor surges after adjustment on MF135 and MF150 with Continental engine, turn bumper screw (S—Fig. 169) in slightly and retest. Do not turn screw farther than is necessary to eliminate objectional surge. On other models recheck carburetor link for proper length. Shorten or lengthen a slight amount from that specified to eliminate surge. Make sure throttle will fully open when adjustment is completed.

If throttle lever creeps or will not maintain a set position, adjust the pressure on friction disc as shown in Fig. 172 or 173.

OVERHAUL

Models MF135 - MF150

151. To overhaul the governor, first remove timing gear cover as outlined in paragraph 45 or 68. The governor lever shaft is supported in needle bearings in timing gear cover. To service the shaft, bearings or oil seals on Continental engine, remove the pipe plug from front of timing gear cover, drive out the groove pin and withdraw the lever. Bearings and seals can be renewed at this time. Procedure for service on Perkins engines is given in paragraph 68.

Governor ball and driver assembly (3—Fig. 175) is available only as a unit which can be renewed after removing camshaft nut (2) on Continental engines or distributor drive shaft nut on Perkins engine. Tighten the camshaft nut to a torque of 70-80 ft.-lbs. or distributor drive shaft nut to 25-30 ft.-lbs.

Withdraw thrust cup assembly (1), make sure the lip tab (L) is not broken or shaft is not scored. Also check ball race for channeling. Shaft of thrust cup should have 0.002-0.006 clearance in camshaft bore. Plunger bore in camshaft is vented by a drilling which emerges from shaft just to rear of front cam; governor action will be impeded if vent is plugged.

Fig. 176 — Exploded view of crankshaft mounted governor weight unit used on Model MF165 with Continental engine.

1. Snap rings	5. Governor balls
2. Fork base	6. Ball driver
3. Thrust bearing	7. Outer race
4. Inner race	8. Thrust Washer

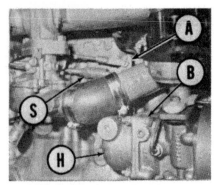

Fig. 177—Installed view of engine governor on Model MF165 tractor equipped with Perkins gasoline engine.

A. Governor arm
B. Governor body
H. Housing
S. Spring

Model MF165 With Continental Engine

152. To overhaul the governor, first remove timing gear cover as outlined in paragraph 45. The governor lever shaft is supported in needle bearings in timing gear cover. To service the shaft, bearings or oil seals, remove lock screw from governor fork and withdraw the lever. Bearings and seals can be renewed at this time.

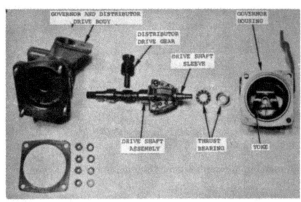

Fig. 178 — Disassembled view of governor and distributor drive assembly used on four cylinder models.

Remove the Woodruff key and withdraw the governor weight unit from crankshaft. Refer to Fig. 176.

To disassemble the governor weight unit, remove the snap rings (1). Renew the governor balls (5), the inner race (4) or outer race (7) if they are worn, scored or discolored.

When installing the timing gear cover, make certain the governor fork falls behind the governor weight unit.

Model MF165 With Perkins Engine

153. Governor arm shaft, bearings and associated parts can be inspected and/or overhauled by unbolting and removing governor housing (H—Fig. 177). To remove shaft and weight unit, it is first necessary to remove timing gear cover as outlined in paragraph 69 and governor gear as in paragraph 73. Shaft and weight unit are only available as an assembly.

Refer to Fig. 178 for a disassembled view of governor unit and to the following data for overhaul:

Distributor drive
shaft clearance 0.0008-0.0024
Distributor drive
shaft end play .....0.004 -0.008
Governor control
shaft clearance 0.0006-0.0023

COOLING SYSTEM

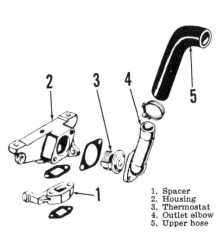

1. Spacer
2. Housing
3. Thermostat
4. Outlet elbow
5. Upper hose

Fig. 179 — Exploded view of thermostat housing and associated parts used on Model MF150 with Continental engine. Model MF135 is similar.

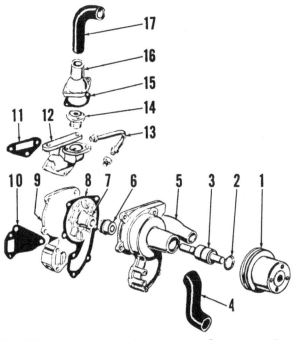

Fig. 180—Exploded view of water pump, thermostat and associated parts used on Model MF165 gasoline tractors with Continental engine.

1. Nut
2. Fan pulley
3. Snap ring
4. Shaft
5. Bearings
6. Spacer
7. Retainer
8. Felt seal
9. Retainer
10. Pump seal
11. Housing
12. Outlet elbow
13. Impeller
14. Bypass elbow
15. Thermostat housing
16. Thermostat
17. Outlet elbow

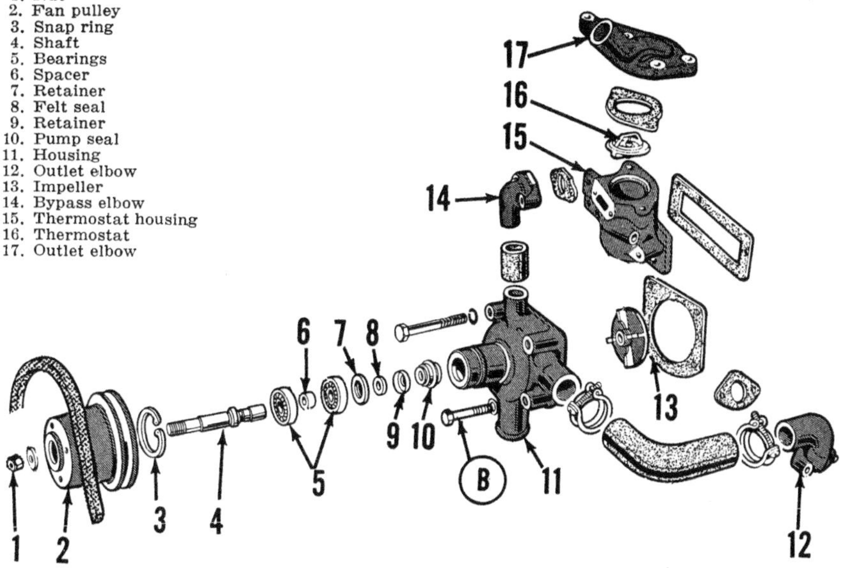

Fig. 181 — Exploded view of water pump, thermostat and associated parts used on Model MF165 diesel. Bolt (B) cannot be withdrawn from housing until pulley (2) has been removed, and must be installed before pulley is installed. If pump is dis-assembled, press bearings (5) and spacer (6) on shaft with open ends of bearings together, then pack the bearings and area between bearings ½-full of high melting point grease before installing shaft in housing. Tighten nut (1) to a torque of 55-60 ft.-lbs.

RADIATOR

All Models

154. All models use a 7 psi pressure type radiator cap. Cooling system capacity is 10 U.S. quarts for MF135 & MF150 gasoline; 10½ U.S. quarts for other models.

To remove the radiator, first drain cooling system and remove hood and grille. Remove air cleaner on Models MF150 and MF165. On all models, disconnect radiator hoses, then unbolt and remove the radiator.

THERMOSTAT

All Models

155. The by-pass type thermostat is contained in a separate housing located behind the outlet elbow. Refer to the appropriate view in Figs. 179 through 182 for exploded view of thermostat and associated parts.

WATER PUMP

All Models

156. Refer to the appropriate exploded views in Figs. 180 through 184 for water pump and associated parts, and to the captions accompanying the views for special overhaul notes.

1. Fan pulley
2. Shaft & bearing
3. Slinger
4. Lower hose
5. Housing
6. Seal
7. Impeller
8. Gasket
9. Cover
10. Gasket
11. Elbow
12. Bypass hose
13. Outlet hose
14. Thermostat housing
15. Gasket
16. Gasket
17. Thermostat
18. Elbow
19. Upper hose

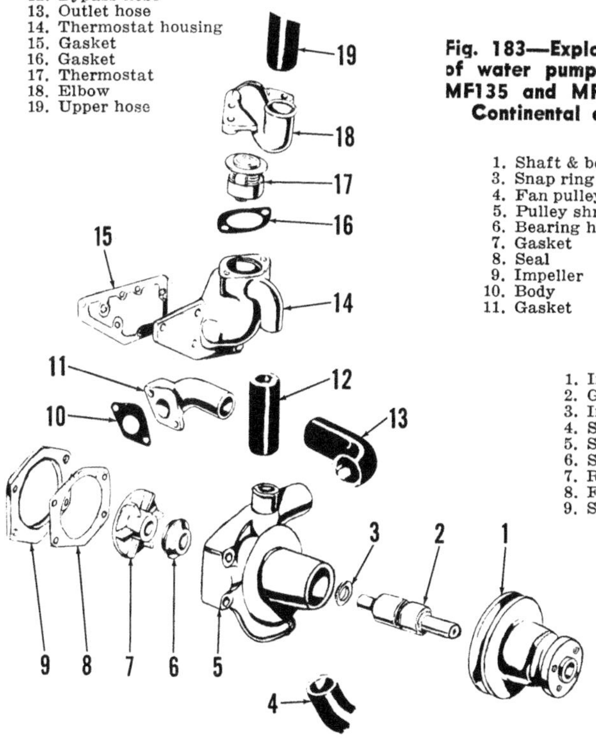

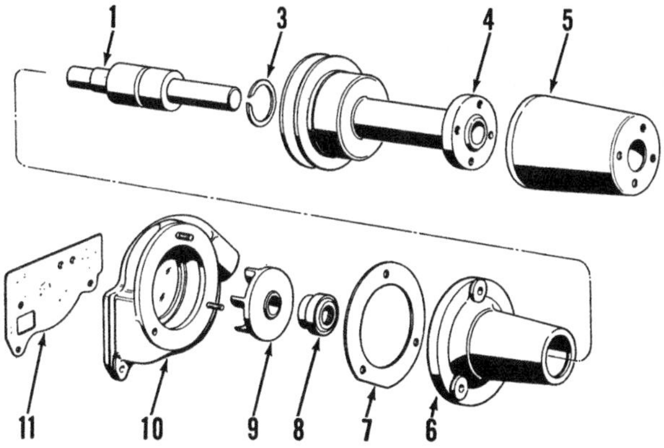

Fig. 183—Exploded view of water pump used on MF135 and MF150 with Continental engine.

1. Shaft & bearing
3. Snap ring
4. Fan pulley
5. Pulley shroud
6. Bearing housing
7. Gasket
8. Seal
9. Impeller
10. Body
11. Gasket

Fig. 182—Exploded view of water pump, thermostat and associated parts of the type used on three cylinder models.

1. Impeller housing
2. Gasket
3. Impeller
4. Shaft
5. Seal
6. Shaft housing
7. Retainer
8. Front seal
9. Seal flange
10. Bearings
11. Spacer
12. Snap ring
13. Fan pulley
14. Nut

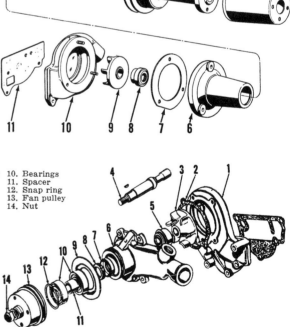

Fig. 184—Exploded view of water pump and associated parts used on Model MF165 tractor with Perkins gasoline engine. Pack the area between bearings (10) half full of high melting-point grease when assembling.

IGNITION AND ELECTRICAL SYSTEM

DISTRIBUTOR

All Models

157. Delco-Remy distributors are used on all models. Refer to paragraph 158 for timing procedure and to paragraph 159 for specifications and overhaul procedure.

158. **TIMING.** On all models except MF165 with Perkins engine, timing marks are located on flywheel, with a timing window on left side of engine block as shown in Fig. 185. On Model MF165 with Perkins gasoline engine, timing marks are on crankshaft pulley (Fig. 185A). A stroboscopic (Power) timing light is recommended for timing adjustment. Initial (static) timing is 6° BTDC for models with Continental engine and 12° BTDC for Perkins engines.

Recommended maximum advance is 28-32 degrees BTDC for MF135 Special with Continental engine and 24-28 degrees BTDC for other models with Continental engine. Maximum advance is 24-28 degrees BTDC for models with three cylinder Perkins engine and 32-36 degrees BTDC for MF165 with Perkins engine.

Firing order is 1-3-4-2 for four cylinder models and 1-2-3 for three cylinder engine. Distributor shaft rotation is counter-clockwise for all models, viewed from rotor end.

159. **OVERHAUL.** Refer to Fig. 186 for an exploded view of the distributor. Centrifugal advance mechanism can be checked for binding or broken

Fig. 185A—On Model MF165 tractor with Perkins gasoline engine, timing marks are located on crankshaft pulley as shown.

springs by turning rotor (21) counterclockwise and releasing, after removing cap (23). Bushings are not available for housing (5); renew housing and/or shaft (8) if clearance is excessive. Shims (3) are available in thicknesses of 0.005-0.010 for adjusting shaft end play which should be 0.002-0.010. Test specifications are as follows:

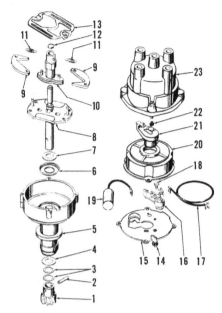

Fig. 186—Exploded view of ignition distributor showing component parts.

1. Drive gear	13. Hold down plate
2. Roll pin	14. Insulator
3. Shims	15. Breaker plate
4. Thrust washer	16. Point set
5. Distributor housing	17. Primary wire
6. Seal	18. Eccentric screw
7. Thrust washer	19. Condenser
8. Shaft assembly	20. Dust shield
9. Advance weight	21. Rotor
10. Cam assembly	22. Brush
11. Spring	23. Distributor cap
12. Oiler wick	

Fig. 185 — View of flywheel timing window used on some gasoline models

Delco-Remy 1112457

Breaker contact gap 0.021
Breaker arm spring tension
   (measured at center of
   contact) ....17-21 oz.
Cam Angle ....45°-48°

Advance data in distributor degrees and distributor rpm

Start advance ....0.0-0.6 @ 250
Maximum advance ......6-8 @ 1000

Delco-Remy 1112458

Breaker contact gap 0.022
Breaker cam spring tension
   (measured at center of
   contact) ....17-21 oz.
Cam Angle ....31°-34°

Advance data in distributor degrees and distributor rpm

Start advance ....0-1 @ 450
Maximum advance .....11-13 @ 1000

Delco-Remy 1112643

Breaker contact gap 0.022
Breaker arm spring tension (measured at center of contact) 17-21 oz.
Cam angle ....31°-34°

Advance data in distributor degrees and distributor rpm

Start advance ....0-2 @ 275
Intermediate advance ......6-8 @ 450
Intermediate advance ....8-10 @ 650
Maximum advance ....11-13 @ 950

Delco-Remy 1112644

Breaker contact gap 0.022
Breaker arm spring tension (measured at center of contact) 17-21 oz.
Cam angle ....31°-34°

Advance data in distributor degrees and distributor rpm

Start advance ....0-2 @ 325
Intermediate advance ......5-7 @ 700
Maximum advance ......9-11 @ 1000

GENERATOR & REGULATOR

All Models So Equipped

160. Delco generators are used on some models. Test specifications are as follows:

Generator

Brush spring tension .... 28 oz.
Field draw
    Volts 12
    Amperes 1.58-1.67
Output (Cold)
    Maximum amperes 25
    Volts 14
    RPM 3040

Regulator

Cutout Relay
    Air gap 0.020
    Point gap 0.020
    Closing voltage (range) .. 11.8-14.0
    Adjust to 12.8
Voltage regulator
    Air gap 0.075
    Voltage range 13.6-14.5
    Adjust to 14.0
Ground polarity .... Negative

ALTERNATOR & REGULATOR

All Models So Equipped

161. **ALTERNATOR.** A "DELCO-TRON" generator (alternator) is used on late models. Units are negative ground.

The only test which can be made without removal and disassembly of alternator is output test. Output should be approximately 32 amperes at 5000 alternator rpm.

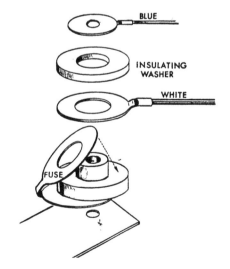

Fig. 187—Exploded view of Alternator output terminal components, showing output fuse/washer which wraps around terminal insulator. Check the fuse if output circuit is dead.

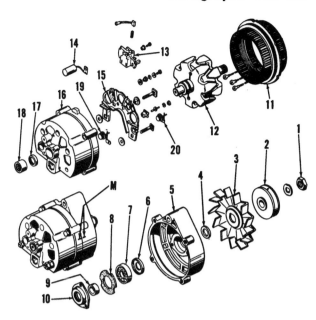

Fig. 188—Exploded view of "DELCOTRON" alternator used on late models.

1. Pulley nut
2. Drive pulley
3. Fan
4. Collar
5. Drive end frame
6. Slinger
7. Bearing
8. Gasket
9. Collar
10. Bearing retainer
11. Stator assembly
12. Rotor assembly
13. Brush holder
14. Capacitor
15. Heat sink
16. Slip ring end frame
17. Felt seal and retainer
18. Needle bearing
19. Negative diode (3 used)
20. Positive diode (3 used)

IMPORTANT: Outlet terminal post of alternator is equipped with a fuse/washer as shown in Fig. 187. If fuse burns out, charging output is cut off from white (output) wire. Current still flows to the blue wire leading to voltage regulator terminal. With charging flow to the battery cut off, voltage rises in the blue (control) wire causing the regulator to cut back the charging current. If generator shows no output, check the fuse before proceeding with further disassembly. Do not attempt to bypass the fuse or connect white wire directly to output terminal. Renew the fuse if damaged, and assemble as shown in Fig. 187.

To disassemble the alternator, first place match marks (M—Fig. 188) on the two frame halves (5 & 16), then remove the four throughbolts. Pry frame apart with a screwdriver between stator frame (11) and drive end frame (5). Stator assembly (11) must remain with slip ring end frame (16) when unit is separated.

NOTE: When frames are separated, brushes will contact rotor shaft at bearing area. Brushes MUST be cleaned of lubricant if they are to be re-used.

Clamp the iron rotor (12) in a protected vise only tight enough to permit loosening of pulley nut (1). Rotor and end frame can be separated after pulley is removed. Check bearing surfaces of rotor shaft for wear or scoring. Examine slip ring surfaces for scoring or wear and windings for overheating or other damage. Check rotor for grounded, shorted or open circuits using an ohmmeter as follows:
Refer to Fig. 189 and touch the ohmmeter probes to points (1-2) and

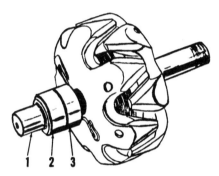

Fig. 189—Removed rotor assembly showing test points when checking for grounds, shorts and opens.

(1-3); a reading near zero will indicate a ground. Touch ohmmeter probes to the two slip rings (2-3); reading should be 4.6-5.5 ohms. A higher reading will indicate an open circuit and a lower reading will indicate a short. If windings are satisfactory, mount rotor in a lathe and check runout at slip rings using a dial indicator. Runout should not exceed 0.002. Slip ring surfaces can be trued if runout is excessive or if surfaces are scored. Finish with 400 grit or finer polishing cloth until scratches or machine marks are removed.

Disconnect the three stator leads and separate stator assembly (11—Fig. 188) from slip ring end frame assembly. Check stator windings for grounded or open circuits as follows: Connect ohmmeter leads successively between each pair of stator leads. Readings should be equal and relatively low. A high reading would indicate an open lead. Connect ohmmeter leads to any stator lead and

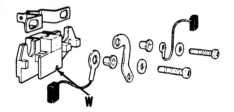

Fig. 190—Exploded view of brush holder assembly. Insert wire in hole (W) to hold brushes up. Refer to text.

stator frame. The three stator leads have a common connection in the center of the windings, a reading other than infinity would indicate a grounded circuit. A short circuit within the stator windings cannot be readily determined by test because of the low resistance of the windings.

Three negative diodes (19) are located in the slip ring end frame (16) and three positive diodes (20) in heat sink (15). Diode should test at or near infinity in one direction when tested with an ohmmeter, and at or near zero when meter leads are reversed. Renew any diode with approximately equal meter readings in both directions. Diodes must be removed and installed using an arbor press or vise and a suitable tool which contacts only the outer edge of the diode. Do not attempt to drive a faulty diode out of end frame or heat sink as shock may cause damage to the other good diodes. If all diodes are being renewed, make certain the positive diodes (marked with red printing) are installed in heat sink and negative diodes (marked with black printing) are installed in end frame.

Brushes are available only in an assembly which includes brush holder (13). Brush springs are available for service and should be renewed if heat damage or corrosion is evident. If brushes are re-used, make sure all grease is removed from surface of brushes before unit is reassembled. When reassembling, install brush springs and brushes in holder, push brushes up against spring pressure and insert a short piece of straight wire through hole (W—Fig. 190) and through end frame (16—Fig. 188) to outside. Withdraw the wire only after alternator is assembled.

Capacitor (14) connects to the heat sink and is grounded to end frame. Capacitor protects the diodes from voltage surges.

Remove and inspect ball bearing (7). If bearing is in satisfactory condition, fill bearing ¼-full with Delco-

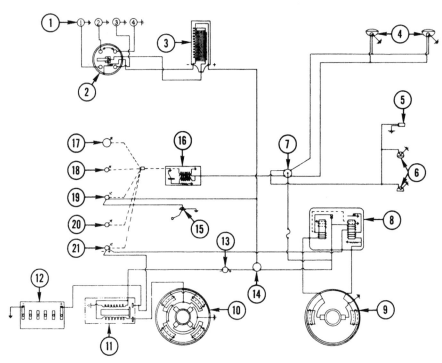

Fig. 191—Wiring diagram typical of that used on early gasoline models.

1. Spark plugs
2. Distributor
3. Ignition coil
4. Headlights
5. Auxiliary light connector
6. Work & warning light
7. Light switch
8. Voltage regulator
9. Generator
10. Starter
11. Starter solenoid
12. Battery
13. Starter safety switch
14. Ignition switch
15. Fuel gage sending unit
16. Power supply
17. Tachometer
18. Oil pressure gage
19. Fuel gage
20. Temperature gage
21. Ammeter

1. Tachometer
2. Oil pressure gage
3. Fuel gage
4. Temperature gage
5. Ammeter
6. Power supply
7. Oil pressure sending unit
8. Fuel gage sending unit
9. Light switch
10. Headlights
11. Auxiliary light connector
12. Work & warning light
13. Voltage regulator
14. Starter safety switch
15. Starter switch
16. Generator
17. Starter
18. Starter solenoid
19. Battery

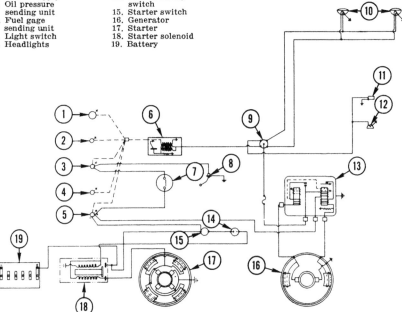

Fig. 192—Wiring diagram typical of that used on early diesel models.

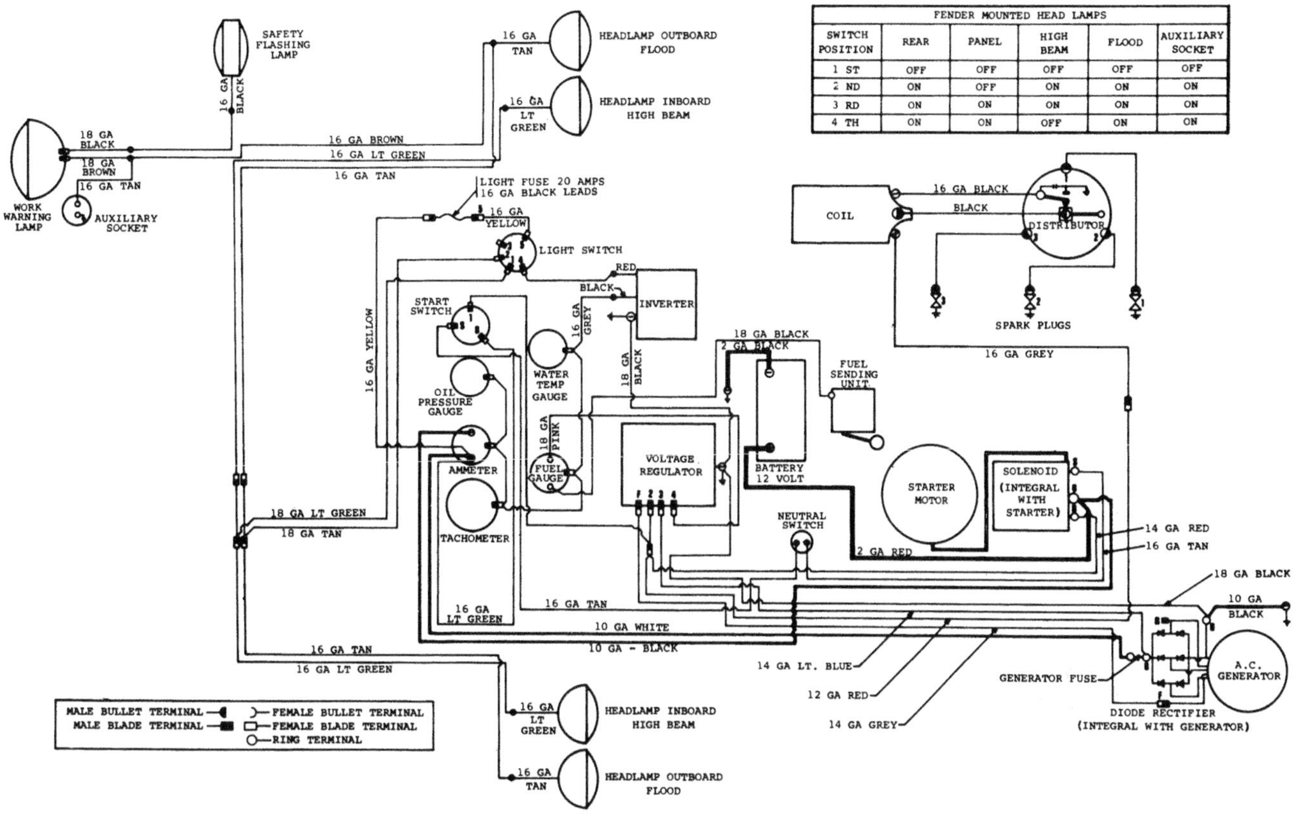

FENDER MOUNTED HEAD LAMPS					
SWITCH POSITION	REAR	PANEL	HIGH BEAM	FLOOD	AUXILIARY SOCKET
1 ST	OFF	OFF	OFF	OFF	OFF
2 ND	ON	OFF	ON	ON	ON
3 RD	ON	ON	ON	ON	ON
4 TH	ON	ON	OFF	ON	ON

Fig. 193—Wiring diagram for MF135 with Perkins gasoline engine. Other late models are similar.

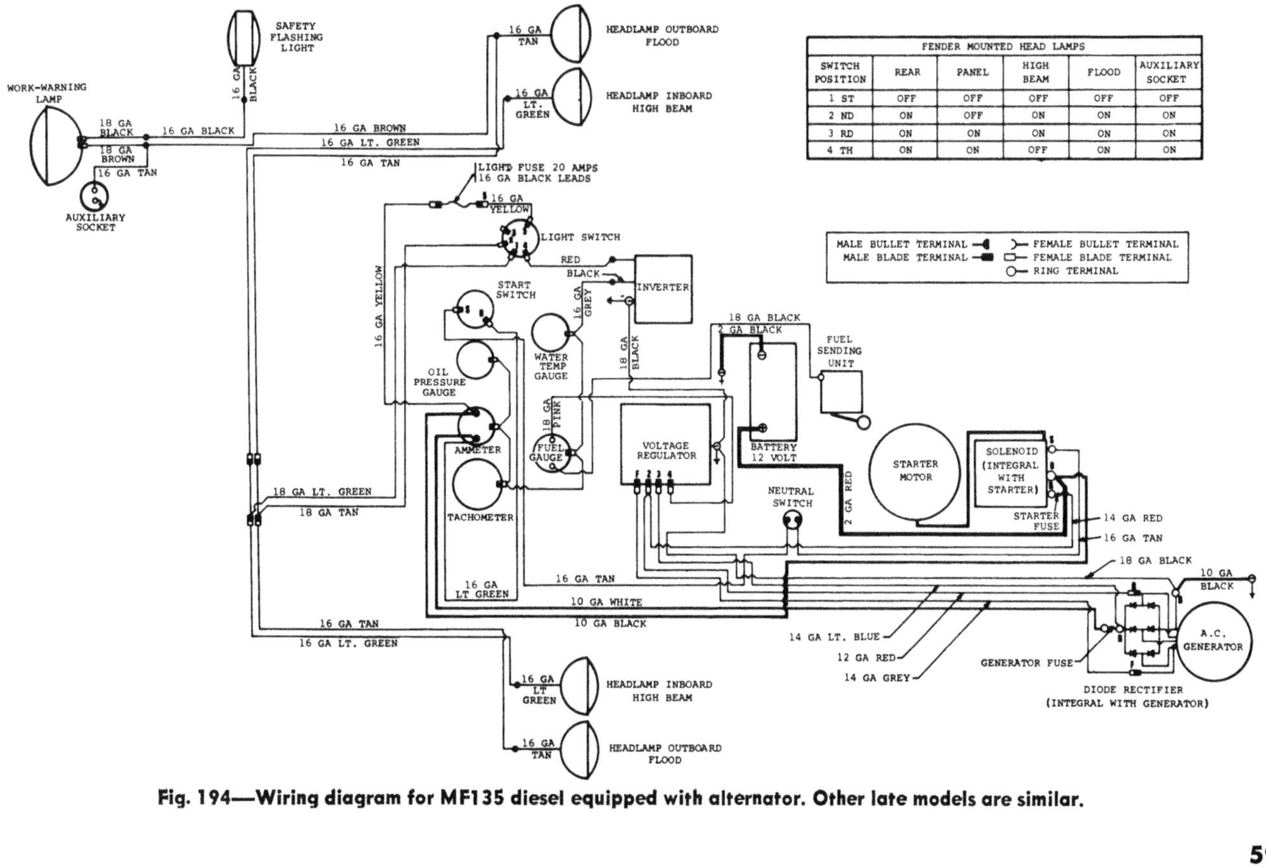

FENDER MOUNTED HEAD LAMPS					
SWITCH POSITION	REAR	PANEL	HIGH BEAM	FLOOD	AUXILIARY SOCKET
1 ST	OFF	OFF	OFF	OFF	OFF
2 ND	ON	OFF	ON	ON	ON
3 RD	ON	ON	ON	ON	ON
4 TH	ON	ON	OFF	ON	ON

Fig. 194—Wiring diagram for MF135 diesel equipped with alternator. Other late models are similar.

Remy Lubricant No. 1960373 and re-install. Inspect needle bearing (18) in slip ring end frame. This bearing should be renewed if its lubricant supply is exhausted; no attempt to re-lubricate the bearing should be made. Press old bearing out toward the in-side and new bearing in from outside until bearing is flush with outside of end frame. Saturate felt seal with SAE 20 oil and install seal and re-tainer assembly.

Reassemble alternator by reversing the disassembly procedure. Tighten pulley nut to a torque of 45 ft.-lbs.

NOTE: A battery powered test light can be used instead of ohmmeter for all elec-trical checks except shorts in rotor winding; however when checking diodes, test light must not be of more than 12 volts.

162. REGULATOR. A Delco-Remy standard two-unit regulator is used. Quick disconnect plugs are used at regulator and alternator. Production regulator is riveted to shock mount; service units are shipped less mount and are attached with screws. Test specifications are as follows:

Regulator Model 1119513

Ground polarity .... Negative
Field Relay
    Air Gap 0.015
    Point Opening 0.030
    Closing Voltage Range 3.8-7.2
Voltage Regulator
    Air Gap 0.067*
    Point Opening 0.014
    Voltage Setting:
      @ 65° F 13.9-15.0
      @ 85° F 13.8-14.8
      @ 105° F 13.7-14.6
      @ 125° F 13.5-14.4
      @ 145° F 13.4-14.2
      @ 165° F 13.2-14.0
      @ 185° F 13.1-13.9

*The specified air gap setting is for bench repair only; make final ad-justments to obtain specified voltage, with lower contacts opening at not more than 0.4 volt less than upper contacts. Temperature (ambient) is measured ½-inch away from regula-tor cover and adjustment should be made only when regulator is at nor-mal operating temperature.

STARTING MOTOR
All Models

163. Delco Remy starting motors are used on all models. Specifications are as follows:

Delco Remy 107329

Brush spring tension .. 35 oz. (min.)
No-Load test
    Volts 10.6
    Amperes (w/solenoid) 49-76
    Minimum rpm 6200

Delco-Remy 1107512

Brush spring tension....35 oz. (min.)
No-Load test
    Volts 10.6
    Amperes (w/solenoid).......75-100
    Minimum rpm 6450

Delco Remy 1108396 or 1108397

Brush spring tension.....35 oz. (min)
No-Load test
    Volts 9
    Amperes (w/solenoid) 50-80
    Minimum rpm 5500

Starter drive pinion clearance is not adjustable, however, some clear-ance must be maintained between end of pinion and starter drive frame, to assure solid contact of the heavy-duty magnetic switch. Normal pinion clearance should be within the limits of 0.010-0.140. Connect a 6-volt bat-tery to solenoid terminals when mea-suring pinion clearance, to keep arma-ture from turning.

ENGINE CLUTCH

Model MF135 Special tractors are equipped with a single disc clutch; all other models use a flywheel mounted dual clutch unit, with two stage pedal control or a split torque clutch and independent power take-off. Refer to the appropriate following para-graphs for adjustment and overhaul proce-dures.

ADJUSTMENT

Model MF135

164. Refer to Fig. 195 for external view of clutch pedal and linkage. To adjust the pedal free play, push down on pedal (P) and measure the clear-ance between pedal arm and radius rod cap at point shown at (C). Clear-ance (C) should be ¾-inch for dual clutch models or 1¼ inches for single clutch models at the moment throwout bearing contacts clutch fin-gers. If it is not, loosen pedal clamp

bolt (B); insert a long punch or rod in hole of throwout shaft (A), turn shaft clockwise until resistance is felt, then reposition pedal on shaft with the specified clearance at (C) and retighten clamp bolt (B).

Fig. 195 — External view of clutch pedal and link-age used on Model MF-135.

A. Throwout shaft
B. Clamp bolt
C. Measure clearance (¾-inch)
P. Clutch pedal

Models MF150 - MF165

165. PEDAL FREE PLAY. Refer to Fig. 196. To check the free play, press down on pedal (P) until resistance is felt, and measure the clearance be-tween release arm stop and transmis-

sion housing at point shown at (C). Free play is correct if clearance (C) measures ⅛-inch. If adjustment is required, insert a long punch or rod in the hole in throwout shaft (A) and loosen the release arm clamp bolt (B). Turn shaft (A) until throwout collar contacts release fingers, reposition release arm with the specified clearance at (C), and retighten clamp bolt (B). After free play has been adjusted, check linkage as outlined in paragraph 166, and adjust if necessary.

166. **LINKAGE ADJUSTMENT.** On dual clutch models. be sure clutch link (D) is in upper hole of release arm (E) as shown in Fig. 196. Place a 1-inch block on step plate beneath clutch pedal pad (P), remove the pto cap, and with pto lever engaged, start the engine. Operate engine at slow idle speed and slowly depress clutch pedal while observing the pto shaft. PTO clutch should release and shaft stop turning at the moment clutch pedal contacts the block on step plate. If adjustment is not correct, adjust the length of link (D).

On all models with single stage clutch, adjust pedal link (D) to a length of 3⅜-inches from center to center of clevis pin holes, then attach clevis to lower hole in release arm (E).

TRACTOR SPLIT
Model MF135

167. To detach engine from transmission assembly, first drain cooling system and remove hood and side panels. Disconnect radius rods and steering drag links at rear end. Shut off fuel and remove fuel line; then, unbolt fuel tank from its rear support and block up between fuel tank and rocker arm cover. Disconnect oil gage line, tachometer cable and heat indi-

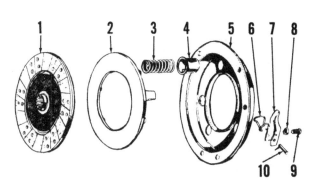

Fig. 197—Exploded view of single clutch used on Model MF135 Special.

1. Clutch disc
2. Pressure plate
3. Clutch spring
4. Spring cup
5. Clutch cover
6. Return spring
7. Release finger
8. Locknut
9. Adjusting screw
10. Pivot pin

cator sending unit. On diesel models, disconnect fuel supply and bleed back lines from final filter and throttle and stop controls from injection pump. On gasoline models disconnect choke cable at carburetor and loosen the U-bolt at front end of throttle control rod. Disconnect battery cables and wiring from generator, starter, coil and fuel gage sender.

Support both halves of tractor separately, remove the attaching bolts and separate the tractor.

Models MF150 - MF165

168. To detach engine from transmission assembly, first drain cooling system and remove hood and side panels. Disconnect battery cables and remove the battery. Remove the cap screws securing instrument support and steering housing to transmission case. Disconnect light wires, wires from starter safety switch, and if tractor is so equipped, the "Multipower" shift linkage. Support the steering column from an overhead hoist and place rolling floor jacks under transmission case and oil pan.

Remove the cap screws securing transmission case to engine or flywheel adapter and carefully separate the units until clutch shaft is clear of clutch unit; then lower front of transmission and/or raise rear of engine until steering and support and instrument panel will clear transmission case and linkage, and roll the units apart.

OVERHAUL

Model MF135 Special
(Single Clutch)

169. Refer to Fig. 197 for an exploded view of clutch pressure plate, cover and associated parts. To overhaul the unit, first mark the cover and pressure plate to assure reassembly in the same relative position, place cover in a suitable press and remove release lever adjusting screws (9) and pins (10).

Release the pressure slowly and disassemble the cover unit. Clutch springs (3) consist of three springs color coded lavender and three springs color coded brown. Springs should have a free length of 2-9/16-inches. The lavender springs should test 180 lbs. when compressed to a height of 1-13/16-inches; brown springs should test 150 lbs. Lavender and brown springs should be installed alternately when reassembling the cover.

Examine pressure plate (2) for score marks or burned areas. Pressure plate may be refaced provided not more than 1/16-inch of the surface is removed.

To assemble the clutch, use a surface plate or serviceable flywheel, and 0.340 key stock spacers in place of clutch disc. Adjust clutch fingers to a height of 1-51/64-inches from surface plate.

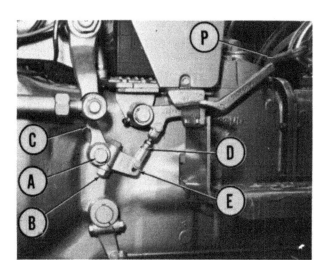

Fig. 196—View of clutch pedal and linkage of the type used on Models MF-150 and MF 165.

A. Throwout shaft
B. Clamp bolt
C. Measure clearance (⅛-inch)
D. Pedal link
E. Release arm
P. Clutch pedal

All Models (Dual Clutch)

170. **REMOVE AND REINSTALL.** To remove the clutch, first split tractor as outlined in paragraph 167 or 168. Make up three special "T" bolts by welding a cross bar to 1/4-20 x 6 inch bolts, then add forcing nuts. Install the "T" bolts in the three holes in outer edge of clutch cover as shown in Fig. 198 and tighten the nuts to compress the springs. Mark the clutch cover (16—Fig. 199), pressure plate (14), drive plate (12) and pressure plate (7), to assure correct assembly and maintain clutch balance; remove the retaining cap screws, and lift the clutch assembly from flywheel as shown in Fig. 198.

To install the dual clutch assembly, first install the air ring using two guide studs as shown. Insert the clutch pilot tool through the driven discs and reinstall by reversing the removal procedure. Tighten the retaining cap screws to a torque of 28-33 ft.-lbs. and adjust, if necessary, as outlined in paragraph 172. Remove the "T" bolts.

171. **OVERHAUL.** To disassemble the removed clutch unit, unhook the three torsion springs (22—Fig. 199) from clutch release levers (17). Back off the lock nuts (8) and thread adjustment screws (11) into pressure plate (7) until they bottom. Back off the forcing nuts on the three "T" bolts until Belleville washer is free and clutch fingers assume the approximate

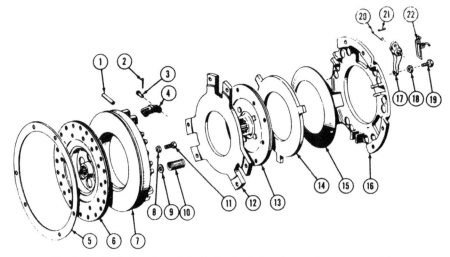

Fig. 199—Exploded view of dual clutch of the type used on most models.

1. Pin	6. Transmission disc	12. Drive plate	18. Lock nut
2. Cotter pin	7. Pressure plate	13. PTO disc	19. Adjusting screw
3. Link pin	8. Lock nut	14. Pressure plate	20. Pivot pin
4. Link	9. Insulating washer	15. Belleville washer	21. Retainer pin
5. Air ring	10. Clutch spring	16. Clutch cover	22. Torsion spring
	11. Adjusting screw	17. Release lever	

position shown in Fig. 200; then drive the groove pins (21—Fig. 199) into the cover until pivot pins (20) can be removed. NOTE: Groove pins can be removed from the bottom after clutch is disassembled. Do not attempt to drive the pins out of clutch cover, or Belleville spring may be damaged.

Remove the pivot pins, back out the forcing nuts on "T" bolts until spring pressure is removed; then remove the

"T" bolts and disassemble the clutch.

Thoroughly clean and examine all parts and renew any which are damaged or worn. Linings are available for both driven plates.

The Belleville spring (15) is color coded red for Models MF135 and MF150; or blue for Model MF165. Coil spring (10) is color coded yellow for Models MF135 and MF150; or lavender for Model MF165. Coil spring specifications are as follows:

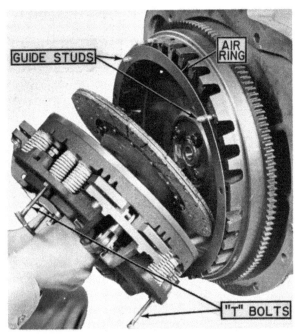

Fig. 198—Special "T" bolts are used for removal and installation of dual clutch. Guide studs are used when installing.

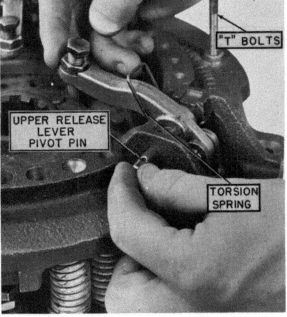

Fig. 200—Removing upper release lever pivot pin.

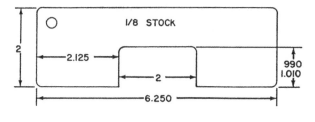

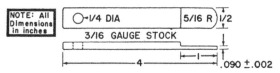

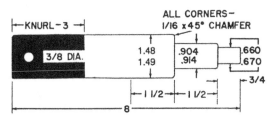

NOTE: All Dimensions in inches

Fig. 201 — Special tools required for dual clutch overhaul and adjustment can be made using the dimensions shown.

Models MF135 - MF150

Free length ....2-45/64 in.

Test length

@ load ....1½ in. @ 80-88 lbs.

Model MF165

Free length ....2-21/64 in.

Test length

@ load ....1½ in. @ 111-123 lbs.

When reassembling the clutch unit, place clutch cover (16) upside down on a bench, then center the Belleville spring (15) in the cover groove with convex side up. Place pressure plate (14) on spring, aligning the previously affixed assembly marks. Install the smaller clutch disc (13) hub down, and drive plate (12) with assembly marks aligned. Temporarily bolt the drive plate and clutch cover together, using three 5/16 x 1½ inch bolts through flywheel mounting holes.

Place the 11-inch pressure plate (7) face down on the bench, coat the lower pivot pins (1) lightly with "Lubriplate" and insert the pins in one ear of pressure plate (7). Position the insulating washers (9) and coil springs (10) on pressure plate and position

the previously assembled clutch cover on the springs with assembly marks aligned. Install the three "T" bolts through holes in clutch cover and threaded holes in pressure plate (7). Tighten the forcing nuts on "T" bolts, compressing the springs until drive plate (12) almost contacts the partially installed pivot pins (1). Install the assembled release levers (17) and links (4) and secure by centering pivot pins (1) in ears of pressure plate.

Tighten the forcing nuts on "T" bolts until pivot pin holes can be aligned in release levers and clutch cover; position the torsion spring (22) and insert the pivot pins (20), making sure positioning holes for groove pins (21) are aligned. Refer to Fig. 200. Use new groove pins (21—Fig. 199) to secure the pivot pins, hook the torsion spring ends, then remove the previously installed 5/16-inch bolts and nuts. Adjust the clutch as outlined in paragraph 172.

172. **ADJUSTMENT.** After the clutch assembly has been installed on flywheel, two adjustments are necessary for proper clutch operation. Proceed as follows:

NOTE: A new 11-inch transmission clutch disc is required when adjusting release lever height. If the removed clutch disc will be re-used, first install the clutch using a new disc, make the adjustment, then assemble using the partially worn parts without changing the adjustment.

Fig. 202 — Adjusting clutch release lever height.

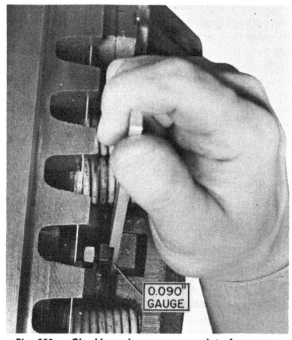

Fig. 203 — Checking primary pressure plate frequency.

Using a finger adjusting tool of the dimensions shown in Fig. 201, adjust the contact surface of all fingers to a height of 1-inch measured from machined rear face of clutch cover as shown in Fig. 202.

With the clutch plate to be used installed, adjust the pto pressure plate clearance to 0.080 for MF165 and 0.090 for other models, as shown in Fig. 203. Adjustment is made by varying the height of adjusting screws (11—Fig. 199) which thread into transmission clutch pressure plate (7).

With both adjustments completed and jam nuts tightened, assemble the tractor as outlined in paragraph 167 or 168, then adjust the linkage as in the appropriate paragraphs 164 through 166.

All Models, Split Torque Clutch

173. Models with independent power take-off use a split torque, single disc clutch with power take-off and hydraulic pump driven by a splined hub carried in clutch cover.

To remove the clutch, first split tractor as outlined in paragraph 167 or 168. Use a suitable aligning tool to install (Fig. 209) and tighten retaining cap screws to a torque of 15-20 ft.-lbs. on Models MF135 and MF150; or 30-35 ft.-lbs. on Model MF165. Long end of clutch plate hub goes to rear on all models.

To disassemble the removed pressure plate and cover unit, place the assembly in a press as shown in Fig. 204 and apply only enough pressure to relieve tension on pins. Remove the pins and lift off cover as shown in Fig. 205 or 206.

Inspect release levers and pins for wear or damage and pressure plate for scoring, heat checks or wear at

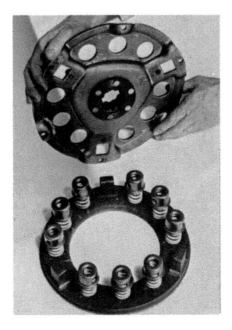

Fig. 205—Removing clutch cover from split torque clutch used on three cylinder models. Note splined drive hub for IPTO drive shaft.

actuating pin holes. Pressure plate may be re-faced if facilities are availble. Inspect pto drive hub in cover for spline wear or looseness. Inspect springs for heat discoloration or other damage. Springs used on Models MF135 or MF150 should test 105 lbs. when compressed to a height of $1\frac{7}{16}$ inches. Springs used on Model MF165 should test 220 lbs. at $1\frac{7}{16}$ inch test height.

Assemble by reversing the disassembly procedure, making sure pins are installed with heads leading in normal rotation as shown in Fig. 204. Clutch fingers should be adjusted to equal height after installation, using adjusting gage MFN202D.

Fig. 206—Removing cover from split torque clutch used on Model MF165 so equipped. Refer also to Fig. 205.

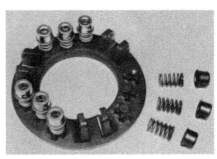

Fig. 207—Nine springs are used in split torque clutch. Refer to text.

Fig. 208—Removing levers from clutch cover used on three cylinder models.

Fig. 204—Using a press to disassemble split torque clutch. Clutch is assembled with head of pivot pin leading as shown.

HEAD OF PIN ON THIS SIDE

Fig. 209—Aligning split torque clutch on Model MF165.

DUAL RANGE TRANSMISSION
(Without Multipower)

TRANSMISSION REMOVAL

All Models

174. To remove the complete transmission unit, first detach (split) engine from clutch housing as outlined in paragraph 167 or 168. Disconnect light wires and unbolt and remove steering gear housing and/or transmission top cover. Drain transmission and rear axle center housing. Disconnect brake rods and remove step plates. Support transmission and rear axle center housings separately, remove the attaching bolts and separate the units. Install by reversing the removal procedure.

TRANSMISSION TOP COVER

All Models

175. **REMOVE AND REINSTALL.** On Model MF135, the transmission top cover is integral with the steering gear housing and removal procedure is outlined in paragraph 13. On other models, remove rear hood side panels and battery. Remove the screws securing instrument panel to transmission cover. Raise instrument panel about 1-inch and secure by blocking, then unbolt and lift off the top cover with shift lever attached.

Overhaul shift levers on top cover as outlined in paragraph 176, install by reversing the removal procedure.

176. **OVERHAUL.** Refer to Fig. 210. To remove either shift lever, compress the spring (2) and unseat and remove the spring seat (1). Pull up the cover (6), drive out the retaining pin (7 or 11); then withdraw shift lever upward out of cover or steering gear housing. Range shift lever (10) contains an "O" ring oil seal (12) which fits against a shoulder in shift lever cup. Shift lever cups (C) are renewable from below in the combined transmission top cover and steering gear housing used on Model MF135, and are machined into cover (4) on other models.

MAIN DRIVE SHAFT
(CLUTCH SHAFT)

All Models

177. **REMOVE AND REINSTALL.** On Model MF135 Special tractor with single clutch, the main drive shaft, gear and housing assembly can be removed from the front after separating clutch housing from engine as outlined in paragraph 167; then removing release bearing and linkage. Refer to Fig. 211 for cross sectional view.

On model with dual or split-torque clutch, it is first necessary to remove the complete transmission assembly as outlined in paragraph 174.

With clutch release linkage disassembled and transmission housing or hydraulic pump removed, refer to Fig. 212 and proceed as follows:

Remove pto bearing front cover plate (27) and unseat and remove snap ring (24) and retaining washer (23). Thread two forcing screws into bearing housing (21) and, tightening the screws evenly, remove bearing housing and associated parts. Bump or pull pto front drive shaft (13) rearward until clear of splines in pto drive gear (19), and lower gear to bottom of transmission housing. Remove the retaining cap screws and withdraw main drive shaft and housing assembly (1 through 12) as a unit from transmission case.

Overhaul the main drive shaft and associated parts as outlined in paragraph 178 and install by reversing the removal procedure.

178. **OVERHAUL.** To disassemble the removed clutch shaft and associated parts, unseat snap ring (11—Fig. 212) and bump transmission drive shaft (12) and bearing (10) rearward out of housing (1). On single clutch models, seal (S—Fig. 211) can be renewed at this time. Install the seal with lip to the rear.

On dual clutch models, unseat and remove snap ring (8—Fig. 212) and bump pto clutch shaft (5) and bearings (4 & 7) from housing. Seal (6)

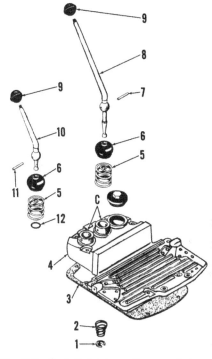

Fig. 210—Exploded view of transmission top cover and shift levers used on Models MF150 and MF165.

1. Spring seat	7. Pin
2. Spring	8. Gear shift lever
3. Gasket	9. Knob
4. Top cover	10. Range shift lever
5. Spring	11. Pin
6. Cover	12. "O" ring

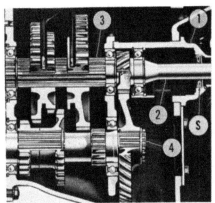

Fig. 211—Cross sectional view of transmission input shaft and gears used on Model MF135 Special with single clutch.

1. Drive shaft housing
2. Main drive shaft
3. Main shaft
4. Countershaft
S. Oil seal

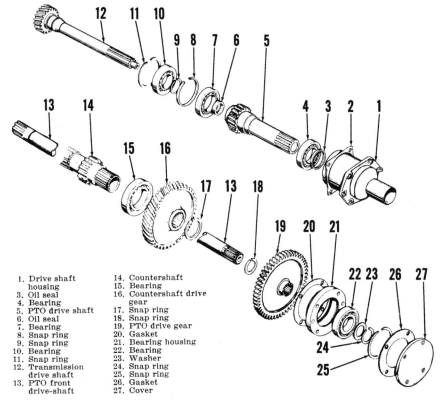

1. Drive shaft housing
3. Oil seal
4. Bearing
5. PTO drive shaft
6. Oil seal
7. Bearing
8. Snap ring
9. Snap ring
10. Bearing
11. Snap ring
12. Transmission drive shaft
13. PTO front drive-shaft
14. Countershaft
15. Bearing
16. Countershaft drive gear
17. Snap ring
18. Snap ring
19. PTO drive gear
20. Gasket
21. Bearing housing
22. Bearing
23. Washer
24. Snap ring
25. Snap ring
26. Gasket
27. Cover

Fig. 212 — Exploded view of transmission input shafts and gears used on dual clutch models without "Multipower" transmission.

to top rear and selector lock grooves to the center.

A change has been made in planetary shift rail (2) and fork (18) which corresponds with a change in the width of the dual range planetary gears. On early models which use the narrow planetary gears, rail (2) is 20-11/16 inches in length and fork (18) is straight. On late models using the wide planetary gears rail (2) is 20-1/32 inches in length and fork (18) is offset to rear when installed. Refer also to paragraph 183 and Fig. 215 and be sure the correct parts are installed if any of the parts are renewed. Tractor serial numbers corresponding to the change are not available.

MAIN (SLIDING GEAR) SHAFT

All Models

180. To remove the transmission main shaft (12—Fig. 214), first remove the transmission assembly as outlined in paragraph 174, clutch shaft as in paragraph 177 and shifter rails and forks as in paragraph 179. Remove the four cap screws securing the planetary unit to rear of transmission case and withdraw rear cover plate (27), thrust washer (19) and planet carrier (26). Using two screw drivers, work the planetary ring gear (18) and dowels from case. Remove planetary front cover (17) and shim (16).

Remove snap ring (9) from front of mainshaft and bump mainshaft rearward out of front bearing; then withdraw shaft (12) and bearing (15) out from rear while lifting gears (11 & 13) out of top opening. To remove front bearing (10) from transmission case, first remove snap ring (39) and slide countershaft drive gear (38) forward off of shaft. Remove main shaft rear bearing (15) forward off of shaft (12) after removing the front snap ring (14). Assemble by reversing the removal procedure.

COUNTERSHAFT

All Models

181. To remove the countershaft (36—Fig. 214), first remove mainshaft as outlined in paragraph 180 and proceed as follows:

Remove snap ring (30) from rear of countershaft and snap ring (39) and gear (38) from front end. Use a suitable step plate in the hollow shaft and bump countershaft (36) forward out of transmission case, bearing (31) and gears (32 & 33). Front bearing

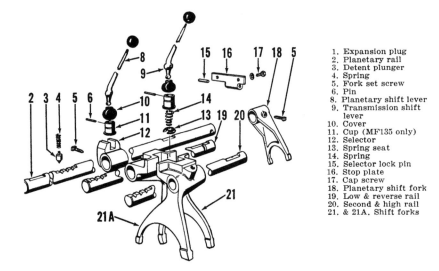

1. Expansion plug
2. Planetary rail
3. Detent plunger
4. Spring
5. Fork set screw
6. Pin
8. Planetary shift lever
9. Transmission shift lever
10. Cover
11. Cup (MF135 only)
12. Selector
13. Spring seat
14. Spring
15. Selector lock pin
16. Stop plate
17. Cap screw
18. Planetary shift fork
19. Low & reverse rail
20. Second & high rail
21. & 21A. Shift forks

Fig. 213 — Exploded view of shifter rails and forks of the type used on all models.

in shaft (5) and seal (3) in housing (1) can be renewed at this time. Install the seals with lips to rear and use seal protecters when reassembling. Assemble by reversing the disassembly procedure and install as outlined in paragraph 177.

SHIFTER RAILS AND FORKS

All Models

179. To remove the shifter rails and forks, first remove transmission top

cover as outlined in paragraph 175 and detach transmission housing from rear axle center housing.

Unwire and remove the set screws retaining the selector and shifter forks to rails. Remove the detent springs (4—Fig. 213), plungers (3) and stop plate (16); then withdraw shifter rails and forks from transmission case.

Forks (21 & 21A) are interchangeable but rails (19 & 20) are not. Rails should be installed with milled flat

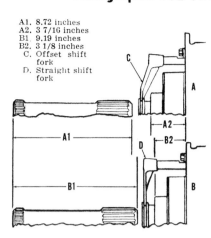

A1. 8.72 inches
A2. 3 7/16 inches
B1. 9.19 inches
B2. 3 1/8 inches
C. Offset shift fork
D. Straight shift fork

Fig. 215—Schematic method of identifying the late (wide) planetary unit (A) and early (narrow) planetary unit (B). Many parts are not interchangeable.

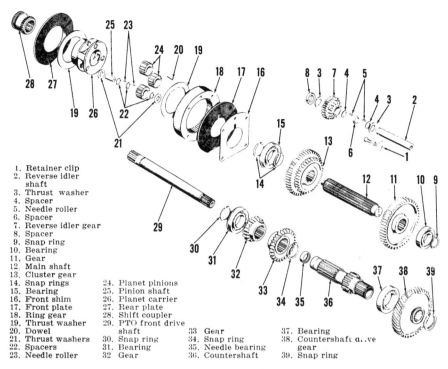

1. Retainer clip
2. Reverse idler shaft
3. Thrust washer
4. Spacer
5. Needle roller
6. Spacer
7. Reverse idler gear
8. Spacer
9. Snap ring
10. Bearing
11. Gear
12. Main shaft
13. Cluster gear
14. Snap rings
15. Bearing
16. Front shim
17. Front plate
18. Ring gear
19. Thrust washer
20. Dowel
21. Thrust washers
22. Spacers
23. Needle roller
24. Planet pinions
25. Pinion shaft
26. Planet carrier
27. Rear plate
28. Shift coupler
29. PTO front drive shaft
30. Snap ring
31. Bearing
32. Gear
33. Gear
34. Snap ring
35. Needle bearing
36. Countershaft
37. Bearing
38. Countershaft drive gear
39. Snap ring

Fig. 214 — Exploded view of transmission shafts, gears and associated parts used on models without "Multipower" transmission.

(37) can be removed from shaft and rear bearing (31) from case at this time. Renew needle bearing (35) in rear of shaft bore if bearing is damaged.

Install by reversing the removal procedure. Use a wooden block or buck up front of shaft by other suitable means while drifting rear bearing (31) on shaft.

REVERSE IDLER ASSEMBLY
All Models

182. The reverse idler gear and shaft can be removed after removing the mainshaft as outlined in paragraph 180, however, removal is not required for removal or installation of countershaft.

Reverse idler shaft (2—Fig. 214) is retained by clip (1) which fits a notch in shaft. Reverse idler bearing consists of two rows of loose needle rollers (5) separated by spacers (4 & 6). A total of 56 rollers (5) are used. Assemble by reversing removal procedure, using Fig. 214 as a guide.

PLANETARY UNIT
All Models

183. The planetary unit can be removed as outlined in paragraph 180, after detaching transmission from rear axle center housing.

Planet pinion shafts (25—Fig. 214) are a tight press fit in planet carrier (26); use a suitable press when re-

moving and installing. Two rows of (27 each) loose needle rollers (23) are used in each planet pinion, separated and spaced by three washers (22). Use viscous grease to stick thrust washers (21) to gears, and completely assemble the bearing, before attempting to install the pinion shafts (25).

Two different planetary units have been used. Refer to Fig. 215. The late (wide) unit (A) and early (narrow) unit (B) can be easily identified by the shape of shifter fork (C or D), or by measuring the distance (A2 or B2) from rear of transmission housing to rear edge of planet carrier. The distance (B2) is 3⅛-inches for early units; or 3-7/16-inches (A2) for late units. Specified length of rear drive shaft is 9.19-inches (B1) for early (narrow) planetary unit or 8.72-inches (A1) for late (wide) unit. Refer also to paragraph 179 for the correct length of planetary shift rail to be used with the respective planetary units.

MULTIPOWER TRANSMISSION

The Massey-Ferguson "Multipower" transmission is a modification of the standard dual-range, three speed transmission, providing an additional hydraulically operated high-low range, thus making available a total of twelve forward and four reverse gear speeds. The "Multipower" unit may be shifted while the tractor is moving under load, without disengaging the transmission clutch. Power for the Multipower" clutch is supplied by a separate hydraulic pump which is mounted in the rear axle center housing on top of the regular tractor hydraulic pump.

OPERATION

All Models So Equipped

184. Refer to Fig. 216. The "Multipower" unit consists of the low-range drive gears (1 & 2), high range drive gears (3 & 4), the hydraulically acuated "Multipower" clutch (5), the

jaw-type overrunning clutch (6), the "Multipower" pump (P) and valve (V).

The two pairs of input gears (1 & 2) and (3 & 4) are constantly meshed. When tractor is operating in low range, power flow is through gears (1 & 2), then through the locked jaw-type clutch (6) to the countershaft. The high-range driven gear (4) is splined to the countershaft and drives the high-range input gear (3) faster than the input gear and shaft (1), the slippage occurring in the released "Multipower" clutch (5). NOTE: The tractor will "Free-Wheel" or coast in Low-Multipower, as the jaw-type clutch will release if countershaft speed exceeds input speed through gear (2).

When tractor is operating in "High-Multipower" range, clutch (5) is hydraulically engaged, locking gear (3) to input shaft and gear (1), and power flow is through gears (3 & 4) to the countershaft. The low-range driven gear (2) continues to be driven by gear (1) at a speed slower than countershaft speed, the slippage occuring at the jaw-type overrunning clutch (6). NOTE: The design of the jaw-type overrunning clutch serves as an automatic hill-holder, preventing tractor from rolling backward when clutch is released with "Multipower" lever in high-range.

CAUTION: A tractor equipped with "Multipower" transmission cannot be parked in gear. ALWAYS set the brakes when tractor is stopped.

ADJUSTMENT

All Models

185. Refer to Fig. 217 for an exploded view of "Multipower" linkage and associated parts. The only adjustment required is adjustment of lower control rod (9) to hand shift lever (3 or 3A).

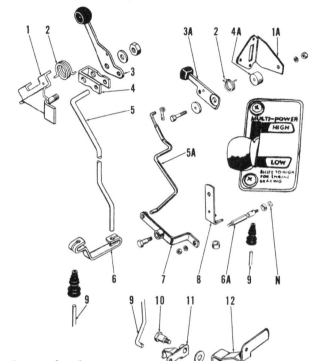

Fig. 217—Exploded view of "Multipower" clutch linkage and associated parts. Model MF135 external linkage is shown on right; other models on left. Internal linkage (9 through 12) is identical for all models.

1. Pivot bracket
1A. Bracket
2. Spring
3. Lever
3A. Lever
4. Spacer
5. Upper link rod
5A. Upper link rod
6. Link
6A. Ball joint
7. Bellcrank
8. Bracket
9. Lower link rod
10. Pivot bolt
11. Shift lever
12. Bracket

To make the adjustment, move hand lever (3 or 3A) fully upward to "High" position. Loosen nut (N) on ball joint (6A) (Model MF135) or clamping bolt in link (6) (Other Models) and push the lower control rod (9) firmly downward as far as it will go. Retighten clamping bolt or nut to lock the adjustment.

TESTING
All "Multipower" Models

186. Tractors are available with the following options relative to the "Multipower" hydraulic circuit:

1. "Multipower" only.
2. "Multipower" plus independent pto.
3. "Multipower" plus auxiliary hydraulic system.
4. "Multipower" plus auxiliary hydraulic system and independent pto.

Type of pump, test procedure and operating pressure depends on the combination being used in the particular tractor being tested.

On tractors without independent pto or auxiliary hydraulics (Group 1

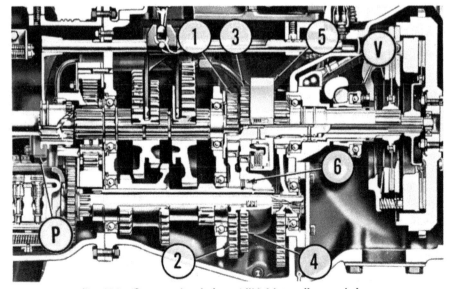

Fig. 216—Cross sectional view of "Multipower" transmission.

P. Multipower pump
V. Multipower valve
1. Low input gear
2. Low driven gear
3. High input gear
4. High driven gear
5. Multipower clutch
6. Jaw clutch

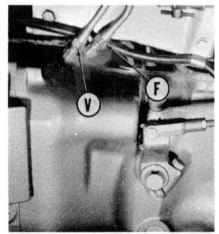

Fig. 218—Multipower oil cooler & filter lines used on some models.

F. Pump to filter   V. Radiator to valve

above), the closed system cannot be coveniently tested. On models with auxiliary hydraulics and without independent pto, a test gauge can be teed into pressure line to filter (F—Fig. 218) to check the "Multipower" system pressure. On models with independent power take-off, a test port is located in pto shift cover in which a gauge can be installed as shown in Fig. 219. At rated engine speed and operating temperature, system pressures should be as follows:

190-200 psi for MF150 Row Crop Diesel and MF165 without independent pto or axiliary hydraulic system.

170-180 psi for MF135 and other MF150 without independent pto or auxiliary hydraulic system.

240-350 psi for early models with independent pto and some models with auxiliary hydraulic system.

200-220 psi for early models with auxiliary hydraulic system.

400-750 psi for late models with Warner-Motive Dual Hydraulic Pump, auxiliary hydraulic system, independent pto and/or Multipower.

Low pressure in "High-Multipower" range could indicate leakage in clutch passages or clutch, improperly adjusted or malfunctioning regulator valve, or malfunctioning pump. Move selector lever to "Low-Multipower" position and recheck.

If equally low pressure is registered in both "Multipower" shift positions, the trouble is either in the pto clutch (if so equipped), "Multipower" control valve, or auxiliary pump. To isolate the control valve as the cause of

trouble, disconnect the line (F—Fig. 218) leading to oil filter and connect a 2000 psi pressure gauge directly into the line leading from transmission case. Start and run tractor ONLY long enough to obtain a pressure reading which should be 500-800 psi. If pressure is as indicated, control valve is at fault; if pressure is low, pump (or independent pto clutch) is faulty.

TRANSMISSION REMOVAL

All Models

187. To remove the complete transmission unit, follow the general procedures outlined in paragraph 174 except disconnect "Multipower" pump supply line before detaching transmission from rear axle center housing.

TRANSMISSION TOP COVER

All Models

188. Refer to paragraph 175 for removal and installation procedure, and to paragraph 176 for overhaul.

CONTROL VALVE

189. **REMOVE AND REINSTALL.** To remove the "Multipower" control valve, first detach (split) engine from transmission housing as outlined in paragraph 167 or 168. Remove clutch release bearing, release fork and shafts, and the brake cross shaft. Refer to Fig. 220. Disconnect shaft linkage (1) and remove shift bracket (3)

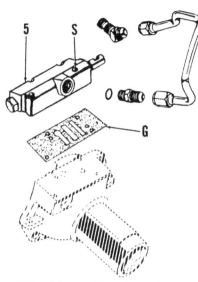

Fig. 221 — Disassembled view of control valve. The bolt which goes in hole (S) retains the valve spool and is sealed with a copper washer. Always use a new gasket (G) when reinstalling valve.

and pressure tube. Remove the cap screws securing the pto input shaft and retainer to front wall of transmission housing and withdraw input shaft, retainer and control valve as a unit from transmission housing.

When detaching control valve (5) from input shaft retainer, note that the cap screw which goes in hole (S—Fig. 221) is sealed with a copper washer. This cap screw also retains the control valve spool; do not withdraw screw from hole (S) when removing or installing the valve.

Always use a new gasket (G) when installing the valve. Make sure the long mounting cap screw with the sealing washer is installed in hole (S) and tighten all screws to a torque of 36-48 inch pounds.

190. **OVERHAUL.** To overhaul the removed "Multipower" control valve, refer to Fig. 222 and proceed as follows:

Remove the sealed cap screw (7) and withdraw control valve spool (9). Remove regulator valve plug (1), spring (2) and valve (3).

Two different regulator valve springs (2) are used on models with "Multipower" only, depending on tractor application. Models MF165 and MF150 Row-Crop Diesel uses a 150 psi spring which is unplated. Other MF150 models and all Model MF135 tractors use a 130 psi spring which is

Fig. 219—Pressure gage can be installed in left side cover on IPTO equipped tractors.

Fig. 220—Front view of transmission housing showing control valve and associated parts.

1. Control rod
2. Actuating lever
3. Bracket
4. Clutch fork
5. Control valve
6. Input shaft

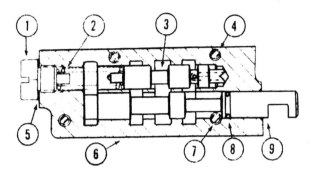

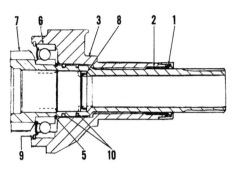

Fig. 222—Cross sectional view of "Multipower" control valve.

1. Regulator plug
2. Regulator spring
3. Regulating valve
4. Mounting cap screw
5. Washer
6. Valve body
7. Long cap screw
8. "O" ring
9. Valve spool

Fig. 224 — Cross sectional view of PTO drive shaft and associated parts. Refer to Fig. 223 for parts identification.

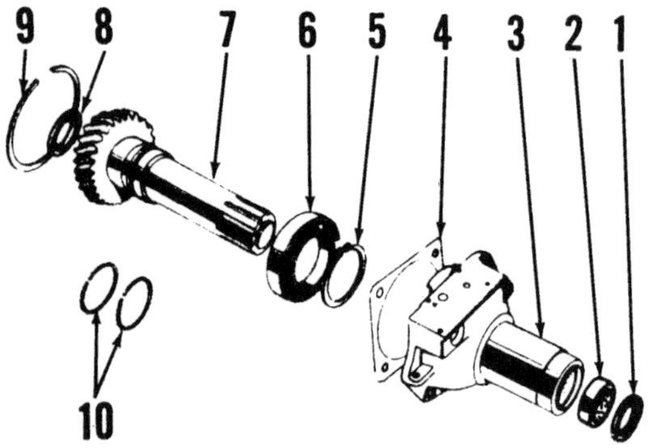

Fig. 223—PTO drive shaft and associated parts used on models with "Multipower" transmission.

1. Oil seal
2. Bushing
3. Housing
4. Gasket
5. Snap ring
6. Bearing
7. PTO drive shaft
8. Oil seal
9. Snap ring
10. Sealing rings

zinc plated for identification. Be sure the correct spring is used if parts are renewed.

After introduction of the Independent Power Take-Off a stronger spring is used, the same spring being used on all models regardless of whether IPTO is installed.

Clean all parts in a suitable solvent, discard "O" ring (8) and carefully examine the parts for wear, scoring or other damage. All parts are available individually.

When reassembling the valve unit, make sure regulating spool (3) is installed with drilled end and identifying notch to closed end of housing bore as shown, and complete the assembly by reversing the disassembly procedure using a new "O" ring (8). Install valve in tractor as outlined in paragraph 189 and adjust the linkage as in paragraph 185.

PTO DRIVE SHAFT AND INPUT SHAFT SEALS

All Models

191. Remove the pto input shaft and housing assembly as outlined in paragraph 189. To disassemble the removed unit, refer to Figs. 223 and 224 and proceed as follows:

Unseat and remove the large snap ring (9) from groove in rear of housing (3), and bump the pto input shaft (7) and bearing (6) rearward

out of housing. Bearing can be removed from the shaft after removing snap ring (5). Seals (1 & 8) are both installed with sealing lips to rear. Inner seal (8) must be carefully positioned in bore of shaft (7), to clear shoulder of transmission input shaft when unit is installed. (Relative installed position of transmission input shaft is indicated by broken lines, Fig. 224). A special removing tool (MFN-850) and installing tool (MFN-849) are available from Massey-Ferguson, Inc. for service on the seal.

Assemble by reversing the disassembly procedure.

Fig. 225—Cross sectional view of transmission main (output) shaft and associated parts.

W. Housing wall
X. Interlock groove
2. Bearing
3. Cluster gear
4. Low gear
5. Main shaft
6. Bearing

SHIFTER RAILS AND FORKS

All Models

192. The transmission shifter rails and forks used on "Multipower" models are identical to those used on standard transmission, and service procedures are covered in paragraph 179.

NOTE: Before transmission can be detached from rear axle center housing, it will first be necessary to remove the pto shift cover, hydraulic lift cover or transmission top cover, and disconnect the "Multipower" oil supply line.

REAR PLANETARY UNIT

All Models

193. The rear planetary unit used on "Multipower" models is identical to standard models, and service procedures are outlined in paragraph 183. Refer also to note under paragraph 192.

TRANSMISSION GEARS & SHAFTS

All Models

194. For service on any of the transmission gears or shafts except pto in-

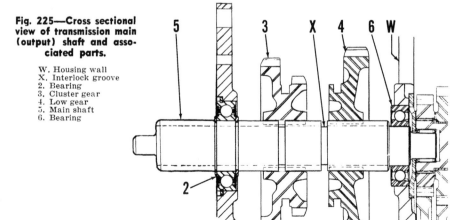

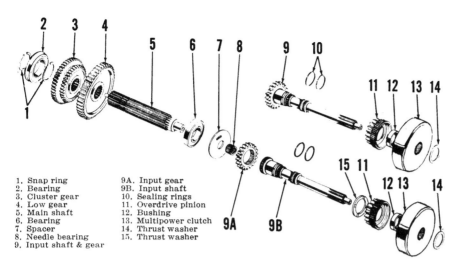

1. Snap ring
2. Bearing
3. Cluster gear
4. Low gear
5. Main shaft
6. Bearing
7. Spacer
8. Needle bearing
9. Input shaft & gear

9A. Input gear
9B. Input shaft
10. Sealing rings
11. Overdrive pinion
12. Bushing
13. Multipower clutch
14. Thrust washer
15. Thrust washer

Fig. 226—Exploded view of transmission upper shafts, gears and associated parts. The two-piece input shaft (9A and 9B) is used on models with planetary final drive.

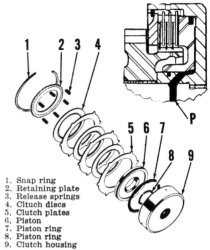

1. Snap ring
2. Retaining plate
3. Release springs
4. Clutch discs
5. Clutch plates
6. Piston
7. Piston ring
8. Piston ring
9. Clutch housing

Fig. 228—Exploded view of "Multipower" clutch and associated parts. Inset shows cross sectional view and pressure passage (P) which enters clutch through the drilled input shafts.

put gear or rear planetary unit, it is first necessary to remove the complete transmission assembly as outlined in paragraph 187, remove the rear planetary unit and shifter rails and forks as previously outlined, then proceed as outlined in the appropriate following paragraphs.

195. **MAIN (OUTPUT) SHAFT.** To remove the main (output) shaft and gears after shift mechanism and rear planetary unit have been removed, refer to Fig. 225 and proceed as follows:

Move the low speed sliding gear (4) forward and insert a large blade screwdriver or similar flat tool in interlock groove (X); then insert a large pry bar between low gear (4) and housing wall (W), and pry the shaft, gears and bearings rearward out of housing bores. Slide the shaft through the gears to bump front bearing (6) from shaft, then remove shaft and rear bearing rearward while lifting the gears out top opening.

When reinstalling, place the cluster gear (3) on shaft with smaller gear to front, and low gear (4) with shift fork groove toward cluster gear. Use a thin spacer plate against housing wall (W) as a support, and bump front bearing (6) into position on shaft; then bump the assembled shaft forward into bearing bores of transmission case.

196. **TRANSMISSION INPUT SHAFT AND "MULTIPOWER" CLUTCH.** To remove the transmission input shaft and "Multipower" clutch unit, first remove the transmission main (output) shaft as outlined in paragraph 195, and the "Multipower" control valve and input housing assembly as in paragraph 189. Move input shaft assembly forward slightly and remove thrust spacer (7—Fig. 226), then withdraw input shaft (9 or 9B) rearward out of case while lifting clutch (13), gears and associated parts out top opening.

Four different input shafts are used. On models which use the planetary final drive (Model MF150 Row-Crop Diesel and all Model MF165) the gear (9A) is too large to pass through bearing bore in housing, so gear is splined to the separate shaft (9B). Different gear ratios are provided for standard and Row-Crop models. Be sure the correct parts are obtained if renewal is indicated.

Different thrust spacers (7) are used with the one-piece input gear (9) and two-piece gear and shaft (9A & 9B). The thrust face which contacts the gear is approximately 7/16-inch wide on the spacer used with one-piece gear (9); and ¾-inch wide on spacer used with gear (9A).

The thrust washer (15) is not used with the one-piece gear and shaft (9).

Input shaft (9 or 9B) contains a needle roller bearing (8) which pilots on front end of main (output) shaft (5), and cast iron sealing rings (10). Examine sealing rings and needle bearing, and inspect the polished bearing and sealing surfaces of shaft for wear or other damage.

Overhaul the removed "Multipower" clutch as outlined in paragraph 197, and assemble by reversing the removal procedure.

197. **"MULTIPOWER" CLUTCH.** To disassemble the removed "Multipower" clutch unit (13—Fig. 226), place unit on a clean bench with overdrive pinion (11) up. Apply slight pressure to clutch retainer plate (2—Fig. 228) and remove snap ring (1) with a narrow blade screwdriver. Completely

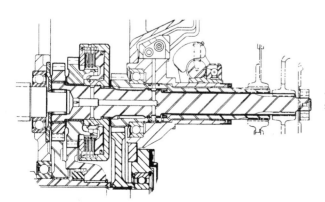

Fig. 227—Cross sectional view of transmission input shaft and "Multipower" clutch assembly.

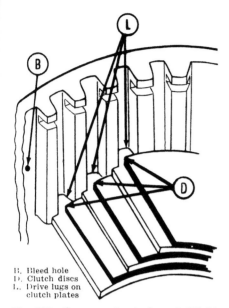

B. Bleed hole
D. Clutch discs
L. Drive lugs on
   clutch plates

Fig. 229—Cross sectional view of "Multipower" clutch showing recommended method of assembly. Refer to text.

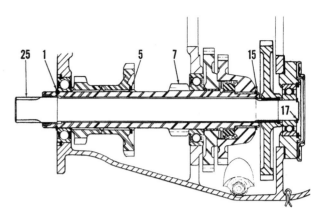

Fig. 230—Cross sectional view of transmission countershaft and jaw-type overrunning clutch. Refer to Fig. 231 for parts identification.

disassemble the clutch and examine the component parts for wear or scoring. Renew the piston sealing rings (7 & 8) whenever clutch is disassembled.

When installing the piston, carefully compress the outer sealing ring (7), using a narrow blade screwdriver or similar tool, and work the piston into its bore. The inner ring (8) will normally compress because of chamfer in inner bore of piston, if care is used in assembly.

When assembling the clutch plates, note that clutch drum (9) contains six bleed holes which are evenly spaced around edge of drum, and that driving plates (5) have six external driving lugs. With piston installed, refer to Fig. 229 and install the first drive plate on top of piston with lugs (L) one spline clockwise from bleed holes (B) as shown. Install an internally splined clutch disc (D), then the second plate with drive lug (L) one spline clockwise from lug of first plate as shown. Install a second clutch disc; then the third plate with driving lugs one spline clockwise from lugs of second plate. Install last plate, then place the piston return springs (3—Fig. 228) on the driving lugs of the first plate installed. Install retainer plate (2) and snap ring (1).

198. COUNTERSHAFT ASSEMBLY. To remove the countershaft (7—Fig. 230), first remove transmission input shaft as outlined in paragraph 196. Remove the cap screws retaining the pto countershaft front bearing plate

(16—Fig. 231) and remove the plate. Remove snap ring (17) and spacer (18) on some early models; or snap ring (17—Fig. 230) on late models. On all models, use two ⅜-inch NC cap screws as forcing screws and remove front bearing retainer (21—Fig. 231) and bearing (20) as a unit. Slide pto front drive shaft (25) out rear of transmission case and countershaft (7), and lift pto drive gear (23) out top opening.

Use the special Nuday clamping tool (MFN-830) or a small "C" clamp to secure the countershaft drive gears (10 and 14) together, then remove snap rings (1 & 15) from each end of countershaft. Insert a step plate of proper size in rear end of countershaft, then bump the shaft forward slightly until snap ring (5) in front

of high speed pinion can be unseated and moved forward on shaft.

With snap ring (5) unseated, insert the step plate in forward end of shaft and bump shaft rearward until the countershaft driving gears (10 & 14) and overrunning clutch (12) can be slipped off forward end of shaft and lifted from transmission.

With drive gears and overrunning clutch removed, drive the countershaft forward until free of rear bearing (2); then slide the shaft and front bearing forward out of transmission case while lifting the gears (3 and 4) out top opening.

Assemble by reversing the disassembly procedure. The two gears (3 & 4) must be installed with hubs together with the larger gear forward.

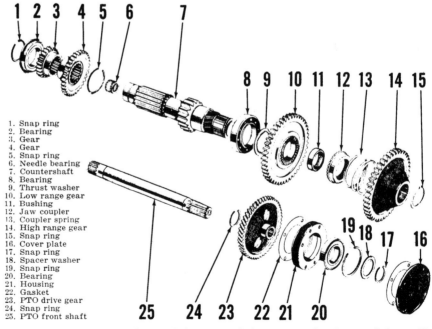

1. Snap ring
2. Bearing
3. Gear
4. Gear
5. Snap ring
6. Needle bearing
7. Countershaft
8. Bearing
9. Thrust washer
10. Low range gear
11. Bushing
12. Jaw coupler
13. Coupler spring
14. High range gear
15. Snap ring
16. Cover plate
17. Snap ring
18. Spacer washer
19. Snap ring
20. Bearing
21. Housing
22. Gasket
23. PTO drive gear
24. Snap ring
25. PTO front shaft

Fig. 231—Exploded view of transmission lower shafts and associated parts. Refer to Fig. 230 for cross sectional assembled view.

The countershaft driving gears (10 & 14) and overrunning clutch assembly may be installed as a unit by using Nuday Tool MFN-830; or individually if the special tool is not used. If the special tool is used, leave clamping screws slightly loose until splines are engaged, then tighten clamp and leave in place until all snap rings are seated, to minimize bouncing of overdrive gear (14) against snap ring.

199. REVERSE IDLER. The reverse idler assembly used in the "Multipower" transmission is identical to that used in the dual range transmission. Refer to paragraph 182 for removal and overhaul procedure.

"MULTIPOWER" PUMP
All Models

Power for the "Multipower" disc clutch is supplied by a separate gear-type pump which mounts on top of the regular hydraulic system pump in rear axle center housing. The "Multipower" pump is driven by a gear (1—Fig. 293) which is attached to the front of the main hydraulic pump camshaft by a cotter pin, and which serves as the pto countershaft coupling.

200. REMOVE AND REINSTALL. Refer to paragraph 251 for removal and installation procedure and paragraph 253 for overhaul procedure of auxiliary pump.

complete left final drive unit or rear axle housing, as outlined in paragraph 210 or 221.

On Model MF135 without differential lock, carrier bearing pre-load is not adjustable. On all models with differential lock, correct carrier bearing pre-load is determined when tractor is assembled, and the correct thickness bearing spacer shield (3—Fig. 233) is selected to provide the proper adjustment. Shield (3) is available in thicknesses of 0.030, 0.035, 0.040, 0.045, 0.050 and 0.055. An 0.005 thick shim is also available which can be used between shield and bearing cup (4) instead of a thicker shield. Differential bearings should be adjusted to approximately zero end play. The recommended method of checking the adjustment if major parts are renewed is by use of Special Tool MFN245Y as shown in Fig. 234. Checking and/or adjustment is not usually necessary unless major parts are renewed.

Main drive bevel gear backlash should be 0.003 - 0.019 and is not adjustable. Backlash can be measured through opening in differential housing wall (Fig. 235) after removing the hydraulic lift top cover.

203. OVERHAUL. To disassemble the removed differential unit, first place correlation marks on both halves of the differential case to insure correct assembly. Remove the eight retaining bolts, lift off the differential lock coupling half (8—Fig. 233) and separate the case halves (9 & 17). Differential pinions (13), spider (12) and side gears (11) can now be removed. Recommended backlash of 0.003-0.008 between the differential pinions and side gears is controlled by the thrust washers (10 & 14).

The main drive bevel ring gear is secured to differential case half (17) with special bolts and nuts, and Grade CV (Blue) LOCTITE is used to retain the nuts. When reinstalling, use LOCTITE of the appropriate grade and tighten nuts to a torque of 110-120 ft.-lbs. Tighten the differential case bolts to a torque of 75-85 ft.-lbs.

DIFFERENTIAL, BEVEL
GEARS AND FINAL DRIVE

Model MF150 Row-Crop Diesel and all MF165 tractors are equipped with planetary final drive units located in outer end of rear axle housings. On all other models, the rear axle shaft splines directly into the differential side gears.

A mechanically actuated, jaw-type differential lock is a factory installed option on Model MF135; standard equipment on

other models. Refer to Fig. 233 for an exploded view of main drive bevel gears, differential and differential lock units.

DIFFERENTIAL
All Models

202. REMOVE AND REINSTALL. The ring gear and differential unit can be removed after removing the

DIFFERENTIAL LOCK
All Models So Equipped

204. OPERATION. The mechanically actuated differential lock assembly is standard equipment on Models MF150 and MF165, and a factory installed option on Model MF135.

When the differential lock foot pedal is depressed, the axle half of

1. Oil seal
2. Carrier plate
3. Spacer shield
4. Bearing cup
5. Fork, differential lock
6. Coupling half
7. Bearing cone
8. Coupling half
9. Differential case half
10. Thrust washers
11. Axle gears
12. Differential spider
13. Differential pinions
14. Thrust washers
15. Pilot bearing
16. Main drive bevel gears
17. Differential case half
18. Bearing cup
19. Bearing cone
20. Carrier plate
21. Bearing cone
22. Bearing sleeve
23. Bearing cone
24. PTO gear hub
26. Adjusting nut
27. PTO gear
28. Snap ring

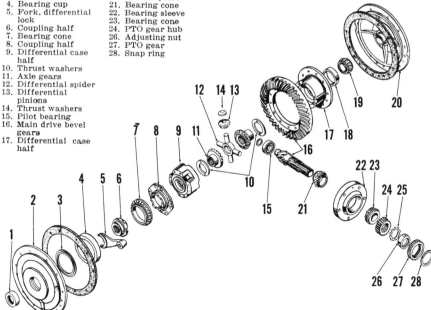

Fig. 233—Exploded view of differential, final drive bevel gears and associated parts. Carrier plates (2 & 20) are not used on models without planetary final drive.

coupler is forced inward to contact the differential case half of the coupler. If slippage is occurring at one wheel, depressing the pedal will cause the coupler dogs to lock the differential case to the right axle. The differential and both drive wheels then rotate together as a unit. As soon as contact is made by the coupler dogs, the pressure will keep the differential lock engaged and foot pedal may be released. When ground traction on both drive wheels again becomes equal, coupler dog contact pressure will be relieved and the coupler will automatically disengage.

Refer to Figs. 236 or 237 for exploded views of the differential lock actuating mechanism and associated parts.

205. **ADJUSTMENT.** The differential lock coupler should be fully engaged when pedal pad (2—Fig. 236 or 237) clears tractor step plate by ¼-inch. If adjustment is required, loosen the clamp screw (4) and reposition pedal (3) on actuating shaft (6—Fig. 236 or 15—Fig. 237). Retighten clamp screw (4) when adjustment is complete.

206. **REMOVE AND REINSTALL.** To remove the differential lock coupler halves (20 & 22—Fig. 236), first drain transmission and hydraulic reservoir, block up under rear axle center housing and remove right fender and rear tire and wheel assembly. Remove right lower hitch link and disconnect right brake linkage. Support

right rear axle housing assembly in a hoist and remove retaining stud nuts; then slide right rear axle and housing as a unit away from rear axle center housing.

Remove bearing cone (21) from coupling half (22); remove the differential case retaining cap screws and lift off the coupling (22). When installing, tighten differential case cap screws to a torque of 75-85 ft.-lbs.

To remove axle half of coupler on models without planetary final drive, unbolt rear axle outer bearing retainer from axle housing (7—Fig. 237) and withdraw axle (6) from housing; then lift out the coupler (8). Shoes (9) will be free in fork (10) with coupler removed. On models with planetary final drive, drive out the roll pin (P—Fig. 238) and remove the two screws securing carrier plate (17) to axle housing; then slip carrier plate (17), fork (19) and sliding coupler (20—Fig. 236) as a unit from axle.

To service the actuating mechanism on models without planetary final drive, remove the clamping bolt (4—Fig. 237) and withdraw actuating shaft (15) from housing (7) and pedal (3), then lift out the remaining actuating parts. On models with planetary final drive, remove cam housing (7—Fig. 236) and brake lever support from axle housing and remove shift fork and coupler as previously outlined. Working through the hole in axle housing, unseat the snap ring (13) and withdraw shaft (16), spring (15) and associated parts.

On all models, assemble by reversing the disassembly procedure and adjust as outlined in paragraph 205 after axle housing is reinstalled.

BEVEL GEARS

All Models

207. **BEVEL PINION.** The main drive bevel pinion is available only in a matched set which also includes the bevel ring gear and attaching bolts and nuts. To remove the bevel pinion, first remove the hydraulic lift cover as outlined in paragraph 247 and proceed as follows:

Working through the top opening in the rear axle center housing, remove the large cotter pin and collapse and remove the rear driveshaft assembly. On models without planetary final drive unit, drain the transmission and hydraulic system fluid and remove left rear axle housing and differential as outlined in paragraph 202.

On all models, unbolt pinion bearing sleeve (22—Fig. 233) from center housing wall and, using two jack screws in the tapped holes, pull the pinion and sleeve assembly forward out of center housing bore.

To disassemble the unit, remove snap ring (28) and gear (27). Unlock and remove nut (26) and bump pinion out of sleeve (22). If bearing cups in sleeve are damaged, renew the sleeve.

When reassembling, tighten nut (26) to obtain a rolling torque of 18-22 inch-pounds for the pinion shaft bearings.

Install the pinion assembly by reversing the removal procedure.

208. **BEVEL RING GEAR.** The main drive bevel ring gear is only avail-

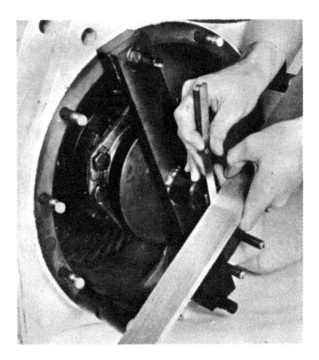

Fig. 234 — Special tools are required for checking differential carrier bearing preload. Refer to text.

Fig. 235—Rear axle center housing with hydraulic lift cover removed. Opening permits inspection of main drive bevel gears and checking gear backlash.

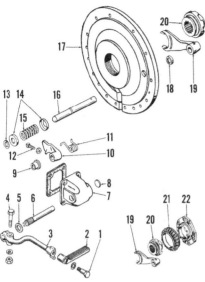

Fig. 238—Installed view of actuating shaft and associated parts used on models with planetary final drive. Refer to Fig. 236 for parts identification.

Fig. 236 — Exploded view of differential lock and associated parts used on models with planetary final drive.

1. Bolt	12. Set screw
2. Pedal pad	13. Snap ring
3. Pedal	14. Spring guide
4. Clamp bolt	15. Release spring
5. Spacer washer	16. Actuating shaft
6. Pedal shaft	17. Carrier plate
7. Cam housing	18. Oil seal
8. Expansion plug	19. Shifter fork
9. Thrust bushing	20. Sliding coupling
10. Actuating cam	21. Bearing cone
11. Return spring	22. Coupling half

able in a matched set which also includes the pinion and attaching bolts and nuts.

To remove the main drive bevel ring gear, first remove the differential assembly as outlined in paragraph 202. Ring gear retaining nuts are installed with LOCTITE, Grade CV, and removal might require a slight amount of heat.

When installing the ring gear, make certain that mating surfaces of ring gear and differential case are thoroughly clean and free from nicks and burrs. Use two drops of LOCTITE, Grade CV (blue) on each attaching bolt and tighten nuts to a torque of 100-120 ft.-lbs. Runout of ring gear should not exceed 0.002.

REAR AXLE AND FINAL DRIVE

Model MF150, except Row Crop Diesel, and all Model MF135 tractors are equipped with a straight drive axle which splines into the differential side gears. A planetary type final drive unit is used on MF150 Row Crop Diesel and all Model MF165 tractors. Refer to the appropriate following paragraphs for removal and overhaul procedures.

Tractors Without Planetary Final Drive

209. BEARING ADJUSTMENT. To check the bearing adjustment, proceed as follows: Support tractor and remove rear tire and wheel assemblies. Shift tractor to neutral, rotate either axle shaft and observe the other axle. If both axles turn in the same direction, the bearing adjustment is too tight. A minimum end play of 0.002 and a maximum of 0.008 should exist in the axle bearing and shafts. End play is controlled by the total shim pack (23—Fig. 239) installed between BOTH axle housing (29) and retainers (35), and varying the thickness of either shim pack will affect the adjustment of both axles. Existing end play may be measured with a dial indicator, or by removing shims from either side until both axles turn in the same direction; then adding shims until clearance exists. Shims (23) are available in thicknesses of 0.004, 0.016 and 0.021.

NOTE: On models equipped with differential lock, do not attempt to make the adjustment by withdrawing RIGHT axle shaft, or sliding differential lock coupler will be lost necessitating draining of the housing and removal of axle sleeve (29).

210. REMOVE AND REINSTALL. Except for the right axle on models equipped with differential lock, axle shaft and bearing only may be removed from axle sleeve (29—Fig. 239) by supporting the tractor, removing wheel and tire unit, and unbolting bearing retainer (35) from axle sleeve (29).

To remove the axle shaft and housing (axle sleeve) as a unit, first drain rear axle center housing, block up and support rear portion of tractor

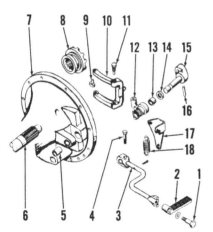

Fig. 237 — Exploded view of differential lock actuating mechanism used on models without planetary final drive.

1. Bolt	10. Actuating fork
2. Pedal pad	11. Pivot bolt
3. Pedal	12. Fork bracket
4. Clamp bolt	13. Bushing
5. Bushing	14. Oil seal
6. Axle shaft	15. Actuating cam
7. Axle housing	16. Groove pin
8. Sliding coupling	17. Spring bracket
9. Coupling shoe	18. Return spring

Fig. 239—Exploded view of rear drive axle and associated parts used on models without planetary final drive.

22. Gasket
23. Shims
24. Bearing cup
25. Expansion plug
26. Gasket
27. Bushing
28. Shaft
29. Axle housing
30. Oil seal
31. Axle shaft
32. Bearing collar
33. Bearing cone
34. Bearing cup
35. Retainer
36. Oil seal

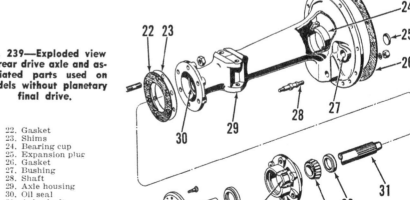

and remove fender and both wheel and tire assemblies. Disconnect brake pedal linkage and lower link. Support axle housing from a hoist and remove the attaching stud nuts and slide the assembly from tractor. Install by reversing the removal procedure, using one gasket (26), to provide the correct differential carrier bearing adjustment.

211. OVERHAUL. Retainer (35—Fig. 239) and bearing cone (33) are secured to axle shaft (31) by a shrink-fit steel collar (32). To renew the axle shaft, bearing cone or cup, retainer or seal (36), first drill the collar (32) and split with a cape chisel; then pull the bearing using a suitable puller or press.

Assemble by reversing the disassembly procedure, seat the bearing cone (33) against shoulder on shaft; then heat a new collar (32) to approximately 800° F. and install and seat against bearing cone before collar cools. Bearing (33) should be properly packed with wheel bearing grease before installation.

Inner oil seal (30) or carrier bearing cup (24) should be inspected and renewed if necessary, while shaft unit is out of axle sleeve.

Models With Planetary Final Drive

Three types of planetary final drive units have been interchangeably used. The type of drive does not affect removal and installation of the complete final drive unit nor removal and installation of wheel axle shaft, drive cover and planet carrier as an assembly. Disassembly, overhaul and adjustment of the planetary unit is affected, however, and the type of unit must be known before parts can be procured.

The two-pinion planetary can be identified by the "184" part number prefix cast into the axle housing (1—Fig. 240) and by the bosses (B) appearing on outside of planetary drive cover (24).

In the early three-pinion (Straddle-Mount) planetary unit, axle housing (1—Fig. 241) carrier a part number prefix of "894 or 899", and a pressed steel shield (23) is attached to inner side of wheel axle flange (24).

In the late three-pinion (Hybrid) planetary unit, axle housing (1—Fig. 242) carries the "184" part number prefix and adjusting shims (S) are visible from the outside.

NOTE: The terms "early" and "late" used in referring the three-pinion planetary units indicate the date first used, not the assembly date of the tractor, as assemblies are used concurrently.

Refer to the appropriate following paragraphs when removing, overhauling or adjusting the units.

212. TWO-PINION PLANETARY UNIT. Refer to paragraph 221 for removal procedure of the complete final drive unit, and to paragraph 222 for removal of wheel axle and planet carrier assembly. Refer to Fig. 243 for an exploded view of the two-pinion planetary unit and associated parts.

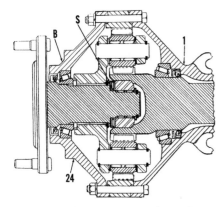

Fig. 240 — Cross sectional view of two pinion planetary drive unit.

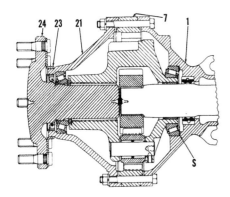

Fig. 241 — Cross sectional view of early three-pinion (straddle mount) planetary final drive unit.

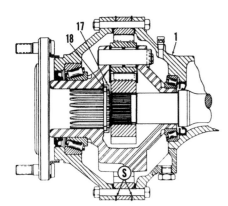

Fig. 242 — Cross sectional view of late three-pinion (hybrid) planetary final drive unit.

1. Axle housing
2. Bearing cup
3. Bearing cone
4. Axle shaft
5. Bearing cup
6. Snap ring
7. Washer
8. Bearing cone
9. Thrust bearing
10. Thrust race
11. Adjusting shims
12. Needle rollers
13. Spacer washers
14. Gasket
15. Ring gear
16. Planet carrier
17. Planet pinion
18. Needle rollers
19. Thrust washer
20. Pinion shaft
21. Segmented ring
22. Bearing cone
23. Bearing cup
24. Drive cover
25. Oil seal
26. Axle sleeve
27. Axle sleeve
28. Wheel axle

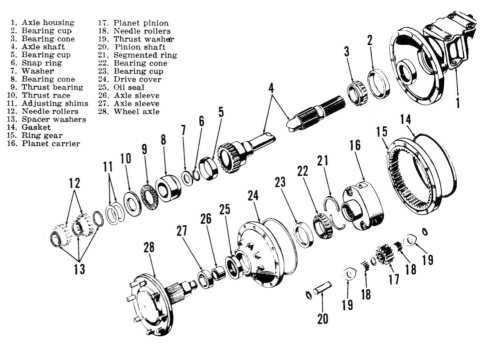

Fig. 243—Exploded view of two-pinion final drive unit used on some tractors. Refer to Fig. 240 for cross sectional view.

213. ADJUSTMENT. Planet carrier bearing preload of 0.001-0.005 is adjusted by varying the thickness of shim pack (11—Fig. 243). This is normally equivalent to a rolling torque of 5-10 ft.-lbs. measured at wheel axle (28) with complete final drive removed as outlined in paragraph 221.

Measure end play with a dial indicator bearing against flange of wheel axle (28) while prying wheel axle out and in using two screwdrivers or pry bars.

If end play exists, record the amount and remove wheel axle and planet carrier assembly as outlined in paragraph 222. Remove snap ring (6) from inner end of wheel axle (28); then lift off washer (7), cone (8), thrust roller (9) and race (10) as a unit, being careful not to lose the loose needle rollers (12) and spacers (13) contained in inner bore of cone. Add shims (11) equal in thickness to the measured end play plus 0.001-0.005, and reassemble by reversing the removal procedure. Shims (11) are available in thicknesses of 0.002, 0.010 and 0.030.

214. OVERHAUL. Left planetary unit can be overhauled after removing wheel axle and planet carrier assembly as outlined in paragraph 222

and locking the brake to permit withdrawal of main axle shaft (4—Fig. 243). If right main axle shaft is to be removed, the complete final drive unit must be removed as outlined in paragraph 221 to correctly install the differential lock sliding coupling. If main axle shaft is not removed, remainder of unit may be overhauled after removing wheel axle and planet carrier assembly as outlined in paragraph 222.

To disassemble the removed wheel axle and planet carrier unit, first remove snap ring (6) and withdraw bearing cone (8) and associated parts, being careful not to lose, damage or intermix the loose needle rollers (12) or shim pack (11); then lift off planet carrier assembly (16). Outer bearing cone (22) is a press fit on wheel axle shaft (28) and thrust load is carried on the split bearing support ring (21); lift out the support ring and press wheel axle out of bearing cone (22) and drive cover (24). Planet pinion shafts (20) can be pressed from planet carrier (16) after removing the retaining snap ring. Two rows of loose needle rollers are used in pinion (17).

When reassembling the unit, place wheel axle (28) on the bed of a press, shaft end up. Install a new seal (25) in drive cover (24) and install drive cover over axle shaft, using appropriate seal protector over shaft splines.

Using a suitable pipe spacer, press bearing cone (22) on shaft only far enough to install the split support ring (21). Install support ring halves with square edge next to bearing cone, then press the shaft back through bearing if necessary, until bearing cone is firmly seated against the support ring.

Install the assembled planet carrier (16) and the assembled cone (8) omitting the shim pack (11). Assemble the wheel axle and planetary unit to axle housing (1) and measure the end play as outlined in paragraph 213, then disassemble only far enough to install the indicated shim pack (11).

215. EARLY THREE-PINION (STRADDLE MOUNT) PLANETARY UNIT. Refer to paragraph 221 for removal procedure of the complete final drive unit, and to paragraph 222 for removal of wheel axle and planet carrier assembly. Refer to Fig. 244 for an exploded view of planetary unit and associated parts.

216. ADJUSTMENT. Planet carrier bearing preload of 0.006-0.015 is adjusted by means of shims (3—Fig. 244) placed in bore of axle housing (1) behind bearing cup (4). Shims are available in thicknesses of 0.005, 0.010 and 0.015. Two methods of adjustment are recommended, a measured adjustment can be made as follows:

Remove wheel axle and planet carrier assembly as outlined in paragraph 222. Remove bearing cup (4) from axle housing (1). Using four bolts of the correct length, bolt planetary ring gear (7) to drive cover (21) as shown in Fig. 245. Support the assembly on drive cover so that wheel axle (24) does not touch, allowing outer carrier bearing (D—Fig. 245) to seat.

Position the removed bearing cup (4) over bearing cone as shown. Place two spacers (S) of known and equal thickness on bearing cup to clear the bearing cage (C); then use a straight edge and inside micrometer to measure the distance (A) from straight edge to surface of ring gear flange. Obtain a measurement from each side of bearing, turn straightedge 90° and remeasure; then add the measurements and divide by four to obtain an average. Subtract the thickness of spacers (S) to obtain the distance from bearing cup to ring gear flange. If the distance is 2.948-2.952, no shims (3—Fig. 244) are required. If distance is less than 2.948, add a sufficient

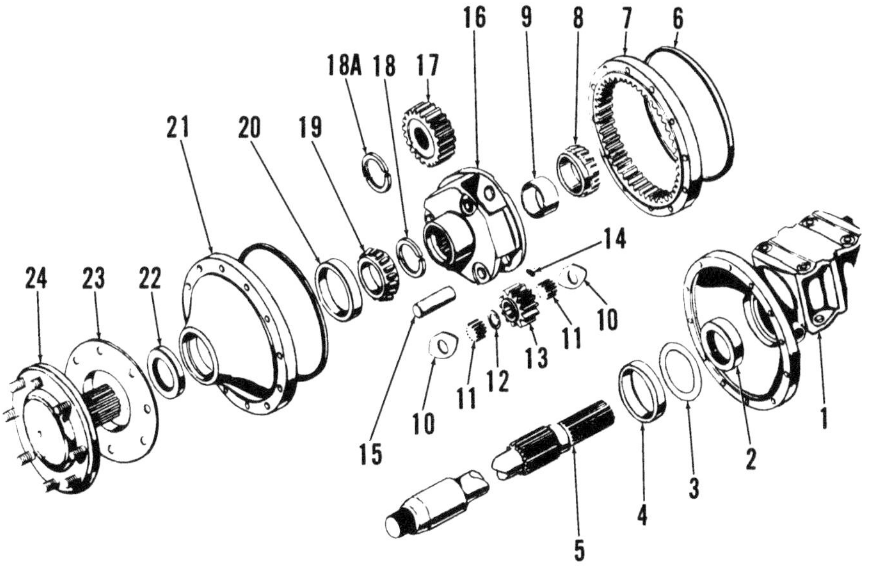

Fig. 244—Exploded view of early three-pinion (straddle mount) planetary final drive unit. Refer to Fig. 241 for cross sectional view.

1. Axle housing	9. Bushing	17. Sun gear
2. Oil seal	10. Thrust washer	18. Snap ring
3. Adjusting shim	11. Needle rollers	18A. Segmented ring
4. Bearing cup	12. Spacer washer	19. Bearing cone
5. Axle shaft	13. Planet pinion	20. Bearing cup
6. Gasket	14. Lock screw	21. Drive cover
7. Ring gear	15. Pinion shaft	22. Oil seal
8. Bearing cone	16. Planet carrier	23. Dust shield
		24. Wheel axle

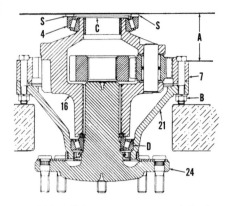

Fig. 245 — Using spacers, a straightedge and inside micrometer to determine bearing adjustment as outlined in paragraph 216.

A. Measure distance	4. Bearing cup
B. Bolts	7. Ring gear
C. Bearing cage	16. Planet carrier
D. Outer bearing	21. Drive cover
S. Spacer	24. Wheel axle

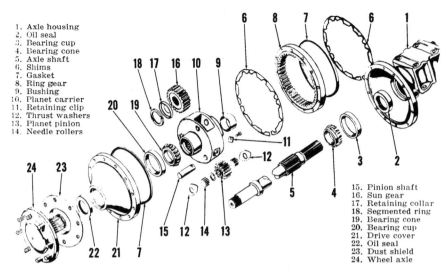

1. Axle housing
2. Oil seal
3. Bearing cup
4. Bearing cone
5. Axle shaft
6. Shims
7. Gasket
8. Ring gear
9. Bushing
10. Planet carrier
11. Retaining clip
12. Thrust washers
13. Planet pinion
14. Needle rollers

15. Pinion shaft
16. Sun gear
17. Retaining collar
18. Segmented ring
19. Bearing cone
20. Bearing cup
21. Drive cover
22. Oil seal
23. Dust shield
24. Wheel axle

Fig. 247—Exploded view of late three-pinion (hybrid) planetary final drive unit. Refer to Fig. 242 for cross sectional view.

quantity or thickness of shims (3) to bring the measurement within the range of 2.948-2.952. Reinstall bearing cup (4) using the proper shims, then reinstall wheel axle and planet carrier assembly as outlined in paragraph 222.

An alternate method of adjustment is to remove the entire final drive unit as outlined in paragraph 221 and stand unit on end with wheel axle up. Remove wheel axle and planet carrier assembly, bearing cup (4) and shim pack (3); then reinstall bearing cup and wheel axle and planetary unit omitting the shim pack. Attach a dial indicator with indicator button contacting wheel axle flange and measure the end play by lifting the wheel axle and planetary unit using to pry bars or screwdrivers. Install a shim pack equal in thickness to the measured end play plus 0.006-0.015, and reassemble by reversing the disassembly procedure.

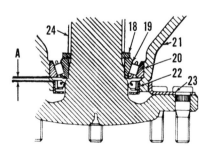

Fig. 246—When installing seal (22), make sure clearance (A) between seal and bearing cup (20) is 0.010. Refer to Fig. 244 for parts identification.

217. OVERHAUL. Left planetary unit can be overhauled after removing wheel axle and planet carrier assembly as outlined in paragraph 222 and locking the brake to permit withdrawal of main axle shaft (5—Fig. 244). If right main axle shaft is to be removed, the complete final drive unit must be removed as outlined in paragraph 221 to correctly install the differential lock sliding coupling. If main axle shaft is not removed, remainder of right unit may be overhauled after removing wheel axle and planet carrier assembly as outlined in paragraph 222.

To disassemble the removed wheel axle and planet carrier unit, loosen locking set screw (14) and remove planet pinion shaft (15), pinion gear and associated parts from largest opening in planet carrier (16); then slide sun gear (17) out the opening.

Wheel axle (24) is a tight press fit in splines of planet carrier (16), and a three leg puller or press of 20-40 tons capacity is required for removal. If a press is used, remove the remaining planet pinions (13) and support the planet carrier. Do not attempt removal by supporting drive cover (21) in the press.

Remove snap ring (18) or ring segments (18A) from axle shaft, then press axle shaft downward out of bearing cone (19).

The seal bore in drive cover (21) is not shouldered. When installing the seal, refer to Fig. 246 and install seal so that 0.010 clearance exists between inner edge of seal (22) and outer edge of bearing cup (20) as shown at (A).

With seal installed, carefully position drive cover (21) over axle shaft and install and seat outer bearing cone (19). Snap ring (18) and split ring (18A—Fig. 244) are interchangeably used and are variable in thickness. The snap ring (18) being available in four different thicknesses from 0.233 to 0.250; and the split ring (18A) available in nine thicknesses from 0.230 to 0.248. With axle and bearing assembled, use the thickest snap ring or split ring which will seat in snap ring groove.

Press the planet carrier on wheel axle shaft using a suitable press, making sure carrier hub seats against snap ring (18). NOTE: When using split ring, make sure segments remain in snap ring groove and properly enter counterbore in carrier hub.

Complete the assembly by reversing the disassembly procedure, adjust carrier bearings as outlined in paragraph 216 and install as in paragraph 221 or 222.

218. LATE THREE-PINION (HYBRID) PLANETARY UNIT. Refer to paragraph 221 for removal procedure of the complete final drive unit, and to paragraph 222 for removal of wheel axle and planet carrier assembly. Refer to Fig. 247 for an exploded view of planetary unit and associated parts.

219. ADJUSTMENT. Planet carrier bearing preload of 0.001-0.005 is adjusted by means of shims (6—Fig. 247) interposed between planetary ring gear (8) and axle housing (1) or cover (21). Shims are available in thicknesses of 0.002, 0.005 and 0.010.

Shims are segmented for easy insertion or removal without removing the retaining bolts, and care must be used to be sure that shim packs are equal in thickness around the complete bolt circle.

To check and/or adjust the preload, first drain the transmission and final drive planetary housing and remove the complete final drive unit as outlined in paragraph 221. Stand the unit on a suitable support, wheel axle up. Loosen the planetary retaining bolts and add 0.010 in shims to the existing shim pack. Attach a dial indicator and measure the existing end play by lifting the wheel axle using two pry bars or screwdrivers; then remove shims equal in thickness to the measured end play plus 0.001-0.005

Fig. 248—Remove retaining clip (11), pinion shaft (15) and planet pinion (13) from largest opening in planet carrier, then remove sun gear.

Fig. 249—Split the retaining collar (17) as shown, using a sharp chisel; then remove retaining collar and segmented rings from wheel axle shaft.

to establish the recommended preload. Tighten the retaining bolts to a torque of 50-55 ft.-lbs.

220. OVERHAUL. To overhaul the planetary final drive unit, first remove the complete final drive as outlined in paragraph 221, remove the retaining bolts and lift off the wheel axle, drive cover and planet carrier as an assembly. Refer to Fig. 247.

Place the removed unit, axle flange down, on a bench. Find the largest planet pinion opening in carrier (10) and remove the cap screw, lock plate (11), shaft (15), pinion (13) and bearings; then withdraw sun gear (16) through the opening. Use a small, sharp chisel and cut the retainer (17) as shown in Fig. 249, then remove the retainer and segmented retaining ring (18—Fig. 247). Using a press or suitable puller, remove wheel axle (24) from carrier (10). The procedure for further disassembly is obvious.

Use a new retainer (17) when reassembling. Heat the retainer to approximately 700° F. and install and seat while hot, then complete the assembly by reversing the disassembly procedure. Adjust the carrier bearing preload as outlined in paragraph 219 after the unit is reassembled.

221. R&R FINAL DRIVE. To remove either final drive assembly as a unit, first drain the transmission, suitably support rear of tractor and remove the fender and rear wheel and tire unit. Disconnect the lower lift link and brake actuating lever. Suitably support final drive assembly from a hoist, remove the attaching stud nuts and separate the final drive unit from rear axle center housing.

Use only one standard gasket when reinstalling, and tighten the retaining stud nuts to a torque of 50-55 ft.-lbs.

222. R&R WHEEL AXLE & PLANET CARRIER ASSEMBLY. To remove the wheel axle and planet carrier as a unit, suitably support tractor and drain the final drive planetary housing. Remove the wheel and tire unit and fender assembly. Securely lock the brake on side to be removed. Remove the securing bolt circle and, using suitable hoisting equipment, move wheel axle and drive cover straight out away from drive axle housing.

Install by reversing the removal procedure. Tighten the retaining bolts to a torque of 50-55 ft.-lbs.

BRAKES

All Model MF135 tractors and MF150 models except Row Crop Diesel are equipped with internal expanding shoe type brakes located at wheel hubs. Model MF150 Row Crop Diesel and all Model MF165 tractors have disc brakes located on main axle shaft at inner ends of final drive housings.

ADJUSTMENT

Shoe-Type Brakes

223. Jack up and block rear portion of tractor. Remove adjusting cover (11—Fig. 251) and, using a screw driver or adjusting tool, turn star wheel (3) by pushing handle of tool toward axle housing until lining contacts brake drum. Back off star wheels an equal amount until drums are free. Equalize the brakes by adjusting the length of rods (15).

Disc-Type Brakes

224. To adjust the disc-type brakes on tractors so equipped, turn each adjusting nut (9—Fig. 252) either way until brake pedal free play is 2½-3 inches when measured at pedal pad. Adjust both brakes equally, then equalize pedal height by means of rods (10).

OVERHAUL

Shoe-Type Brakes

225. The brake drum can be removed after removing the rear wheel and the two drum retaining screws. Remove the positioning spring and pin assemblies (8—Fig. 251), anchor springs (2) and brace (4); then lift off shoes (6), adjuster (3) and return spring (7). Assemble by reversing the

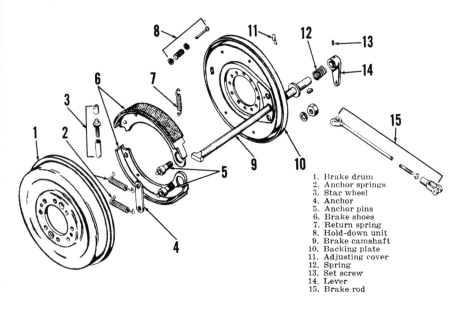

1. Brake drum
2. Anchor springs
3. Star wheel
4. Anchor
5. Anchor pins
6. Brake shoes
7. Return spring
8. Hold-down unit
9. Brake camshaft
10. Backing plate
11. Adjusting cover
12. Spring
13. Set screw
14. Lever
15. Brake rod

Fig. 251—Exploded view of shoe type individual wheel brakes used on models without planetary final drive.

On all models, remove adjusting nut (9—Fig. 252), block (8) and lever support (5); then withdraw lined discs (2) and actuating disc (1) from inner side of axle housing.

Actuating disc assembly (1) can be disassembled after removing brake rod (3) and unhooking the return springs (27). Renew any parts which are questionable and assemble by reversing the disassembly procedure. Adjust the brakes as outlined in paragraph 224.

BELT PULLEY

227. OVERHAUL. To overhaul the belt pulley attachment, refer to Fig. 253 and proceed as follows:

Remove input shaft bearing housing (22) and expansion plug (9). Remove the cotter pin and nut (10) and bump shaft and gear (18) out of housing. Remove pulley (1), nut (2) and hub (5), and bump pinion shaft (17) out of outer bearing and housing.

When reassembling the unit, install pinion shaft (17), bearings and hub (5), and tighten nut (2) until a rolling torque of 2-4 inch pounds is applied to shaft bearings. Secure nut (2) with a cotter pin. Install drive gear (18) and tighten nut (10) until the bevel gears have a backlash of 0.004-0.006; then back off the nut if necessary, until cotter pin can be inserted. Fill to level of filler plug with SAE 90 gear oil.

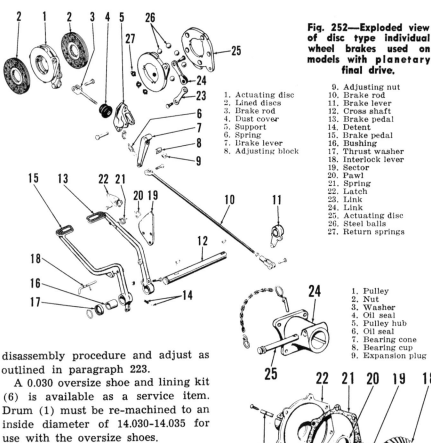

Fig. 252—Exploded view of disc type individual wheel brakes used on models with planetary final drive.

1. Actuating disc
2. Lined discs
3. Brake rod
4. Dust cover
5. Support
6. Spring
7. Brake lever
8. Adjusting block
9. Adjusting nut
10. Brake rod
11. Brake lever
12. Cross shaft
13. Brake pedal
14. Detent
15. Brake pedal
16. Bushing
17. Thrust washer
18. Interlock lever
19. Sector
20. Pawl
21. Spring
22. Latch
23. Link
24. Link
25. Actuating disc
26. Steel balls
27. Return springs

disassembly procedure and adjust as outlined in paragraph 223.

A 0.030 oversize shoe and lining kit (6) is available as a service item. Drum (1) must be re-machined to an inside diameter of 14.030-14.035 for use with the oversize shoes.

Disc-Type Brakes

226. To remove the brake assemblies, first remove differential lock (right side) as outlined in paragraph 206, or complete final drive unit (left side) as outlined in paragraph 221. On left side, remove the retaining screws and lift off the differential carrier plate.

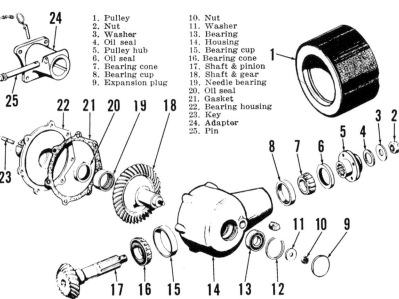

1. Pulley
2. Nut
3. Washer
4. Oil seal
5. Pulley hub
6. Oil seal
7. Bearing cone
8. Bearing cup
9. Expansion plug
10. Nut
11. Washer
13. Bearing
14. Housing
15. Bearing cup
16. Bearing cone
17. Shaft & pinion
18. Shaft & gear
19. Needle bearing
20. Oil seal
21. Gasket
22. Bearing housing
23. Key
24. Adapter
25. Pin

Fig. 253—Exploded view of pto mounted belt pulley attachment available for all models.

POWER TAKE-OFF

(Constant Running & Transmission Driven)

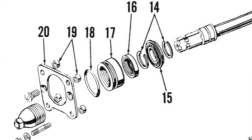

OUTPUT SHAFT

All Models

228. To remove the pto output shaft, drain the transmission and hydraulic system fluid and unbolt rear bearing retainer (20—Fig. 254) from rear axle center housing. Withdraw shaft (13), rear bearing (15) and seal housing (17) as an assembly.

Rear bearing (15), seal (16) or sealing "O" ring (18) can be renewed at this time, as can shaft (13) or seal retainer (17).

Output shaft front support (10) and associated parts can be removed for service of bushing (9) or needle bearing (11) after first removing hydraulic pump as outlined in paragraph 251 and differential unit as outlined in paragraph 202. Use a step plate and driver of the appropriate size to just pass through bushing (9) and drive needle bearing (11) and retainer (12) as a unit out to rear; then drive retainer (10) and bushing (9) forward out of housing wall. The pre-sized bushing (9) is flanged at front end and flange should be seated when installing. Install needle bearing (11) flush with edge of retainer (12). Assemble by reversing the removal procedure, making sure that holes in retainer (12) are at top and bottom when installed.

GROUND SPEED GEARS

All Models So Equipped

229. To remove the ground speed drive gear, first remove the hydraulic lift cover as outlined in paragraph 247 and proceed as follows:

Working through the top opening in rear axle center housing, collapse and remove the rear drive shaft assembly. Remove the retaining snap ring and slide the gear forward off of drive pinion.

To remove the driven gear (8—Fig. 254) after drive gear is out, remove the hydraulic pump as outlined in paragraph 251, remove pto shift cover (4) and slide the gear forward out of bushing (9). Assemble by reversing the disassembly procedure.

PTO MAIN DRIVE GEARS

All Models

230. The pto main drive gears and clutch are included in the transmission drive train. Refer to the appropriate transmission and clutch sections for removal and overhaul.

1. Shift lever	10. Sleeve
2. Washer	11. Needle bearing
3. Seal	12. Retainer
4. Shift cover	13. Output shaft
5. Spring	14. Snap rings
6. Detent	15. Bearing
7. Shift fork	16. Oil seal
8. Gear & coupling	17. Seal retainer
8A. Shift coupling	18. "O" ring
9. Bushing	19. Spacers
	20. Retainer

Fig. 254—Exploded view of pto output shaft and associated parts.

INDEPENDENT POWER TAKE-OFF

Tractors may be optionally factory equipped with an Independent Power Take-Off which is driven by a flywheel mounted "Split-Torque" clutch and controlled by a hydraulically actuated multiple disc clutch contained in rear axle center housing.

provided by the auxiliary gear type pump which also provides power for the Multipower clutch and/or auxiliary hydraulic system if tractor is so equipped.

OPERATION

All Models So Equipped

231. On models with Independent Power Take-Off, the pto drive shaft is splined into a hub contained in the Split Torque Clutch Cover (paragraph 173) and turns continuously when engine is running.

The IPTO control lever is mounted on left side cover of rear axle center housing as shown in Fig. 255. Moving lever forward disengages the hydraulically actuated multiple disc clutch and engages the brake. Moving lever rearward releases the hydraulic brake and engages the multiple disc pto clutch.

Standby hydraulic pressure to actuate the hydraulic clutch and brake is

1. Control lever
2. Pressure line
3. Multiple disc clutch
4. Retainer sleeve
5. Auxiliary pump
6. Multipower pressure holes

Fig. 255—Schematic view of IPTO clutch and associated parts used on some models.

Fig. 256—IPTO output shaft showing component parts.

1. Cap
2. Retainer
3. O-ring
4. Seal retainer
5. Oil seal
6. Output bearing
7. Snap ring
8. Output shaft
9. Needle bearing
10. Snap ring

REMOVE AND REINSTALL
All Models So Equipped

232. **OUTPUT SHAFT.** To remove the IPTO output shaft, first drain transmission & hydraulic system fluid then unbolt retainer plate (2—Fig. 256) from rear axle center housing. Remove protective cap (1). Insert a punch or similar tool in hole in hole in shaft (8) and withdraw shaft, bearing, retainer and associated parts.

Rear bearing (6), seal (5) or sealing O-ring (3) can be renewed at this time. Front needle bearing (9) is contained in center housing wall and renewal requires splitting tractor and removing clutch pack as outlined in paragraph 233.

Install by reversing the removal procedure. Apply pressure to seal retainer (4) as it enters housing, to prevent shaft being pushed forward out of retainer and damaging seal lip.

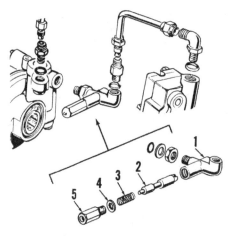

Fig. 258—Adjust clutch clearance after installation by moving retainer sleeve (4—Fig. 255).

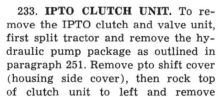

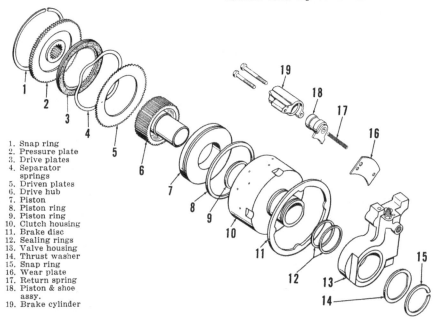

Fig. 259—Exploded view of regulating valve used on IPTO equipped models without Multipower.

233. **IPTO CLUTCH UNIT.** To remove the IPTO clutch and valve unit, first split tractor and remove the hydraulic pump package as outlined in paragraph 251. Remove pto shift cover (housing side cover), then rock top of clutch unit to left and remove through center housing front opening.

Reinstall by reversing the removal procedure. After hydraulic pumps are installed, loosen set screw in retainer sleeve (4—Fig. 255) and adjust clearance between sleeve and clutch hub to 0.005-0.015 by moving the sleeve forward or rearward as required. See Fig. 258. Tighten the set screw after adjustment is obtained.

234. **IPTO REGULATING VALVE.** On "Multipower" transmission models, the IPTO regulating valve is combined with the "Multipower" Control Valve and removal procedure is contained in paragraph 189. On models without "Multipower" transmission, the regulating valve is installed in pressure line leading from pump to clutch valve as shown in Fig. 259. To remove or service the valve, remove hydraulic lift cover as outlined in paragraph 247.

OVERHAUL

All Models So Equipped

235. **OUTPUT SHAFT.** With output shaft removed from tractor as outlined in paragraph 232, seal retainer (4—Fig. 256) can be withdrawn toward rear. Seal (5) bottoms against shoulder in retainer bore and is installed with lip to front.

Fig. 257—Installed view of IPTO clutch with pump package removed.

1. Snap ring
2. Pressure plate
3. Drive plates
4. Separator springs
5. Driven plates
6. Drive hub
7. Piston
8. Piston ring
9. Piston ring
10. Clutch housing
11. Brake disc
12. Sealing rings
13. Valve housing
14. Thrust washer
15. Snap ring
16. Wear plate
17. Return spring
18. Piston & shoe assy.
19. Brake cylinder

Fig. 260—Exploded view of IPTO multiple disc clutch and hydraulic brake unit.

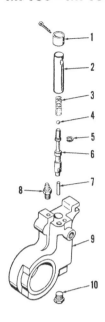

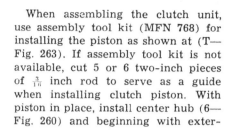

Rear bearing (6) is a press fit on shaft and is secured by retaining ring (7).

236. **REGULATING VALVE.** Fig. 259 shows an exploded view of IPTO regulating valve used on models without "Multipower" transmission. Plug (5) is strongly spring loaded and should be removed with care if disassembly is attempted. All parts are available individually except valve body (1). If body is damaged, renew valve assembly.

On models with "Multipower" transmission, IPTO regulating valve is contained in the "Multipower" control valve and overhaul procedure is contained in paragraph 190.

237. **IPTO CLUTCH & VALVE UNIT.** To disassemble the removed IPTO clutch and valve unit, place the assembly on a bench, valve (front) side down. Remove snap ring (1—Fig. 260), drive plates (3), separator springs (4) and driven plates (5). Lift out drive hub (6). Using two pairs of pliers, grasp strengthening ribs of piston (7) and lift out piston.

Unseat and remove snap ring (15) and lift off valve housing (13) and associated parts. Brake disc (11) will be removed with valve housing. Brake cylinder and associated parts can be disassembled after removing the retaining cap screws.

To disassemble the modulating valve (parts 1 through 6—Fig. 261) apply slight pressure while unseating the internal expanding snap ring (5). Pin (7) is used in early models only.

Spacer ball (4) is available in alternate diameters of $\frac{1}{4}$, $\frac{9}{32}$, $\frac{5}{16}$ and $\frac{11}{32}$ inch and is used to establish collapsed length of modulating valve within the recommended 4.030-4.060 as shown in Fig. 262.

When assembling the clutch unit, use assembly tool kit (MFN 768) for installing the piston as shown at (T—Fig. 263). If assembly tool kit is not available, cut 5 or 6 two-inch pieces of $\frac{3}{16}$ inch rod to serve as a guide when installing clutch piston. With piston in place, install center hub (6—Fig. 260) and beginning with exter-

1. Retainer
2. Plunger
3. Spring
4. Spacer ball
5. Retaining ring
6. Valve spool
7. Pin
8. Connector
9. Housing
10. Plug

Fig. 261—Exploded view of modulating valve, valve housing and associated parts.

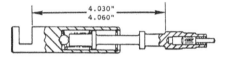

Fig. 262—Collapsed length of modulating valve should be 4.030—4.060 as shown. Length is adjusted by installing a different size spacer ball (4—Fig. 261).

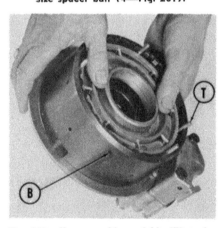

Fig. 263—Use assembly tool kit (T) to install piston as shown. Oil bleed holes (B) are used in assembly as outlined in text and shown in Fig. 264.

Fig. 264—Insert Allen wrenches or rod ends in bleed holes (B—Fig. 263) to compress the separator springs during clutch assembly. Refer to text.

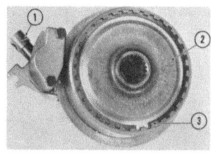

Fig. 266—Rear view of assembled IPTO clutch unit.

1. Cotter pin
2. Pressure plate
3. Snap ring

Fig. 267—Exploded view of IPTO shift cover showing lever components and detent plunger.

Fig. 265—Front view of assembled IPTO clutch unit.

1. Modulating valve
2. Snap ring
3. Thrust washer
4. Clutch housing
5. Brake disc

Fig. 268—Move IPTO control lever forward to disengage IPTO clutch and rearward to engage clutch.

nally splined separator plate (5), alternately install seven separator plates (5), six wave springs (4) and six friction discs (3). Push down on the last separator plate to compress the wave springs and insert two Allen wrenches (or rod ends) in bleed holes (B—Fig. 263) as shown in Fig. 264, to hold the plate in place. Install the remaining friction disc and wave spring, then install pressure plate (2—Fig. 260) and snap ring (1). The restraining Allen wrenches can be removed at this time.

NOTE: If new friction discs are installed, they should be soaked for 30 minutes in transmission and hydraulic fluid before installation.

Install the assembled clutch unit as outlined in paragraph 233.

238. **HYDRAULIC PUMP.** Power for the IPTO clutch and brake is supplied by the gear type auxiliary pump which also supplies the "Multipower" transmission unit and/or auxiliary hydraulic system if so equipped. Refer to paragraph 253 for overhaul procedures.

HYDRAULIC SYSTEM

The hydraulic system consists of a pto driven piston type pump which is submerged in the operating fluid; and a single acting ram cylinder enclosed in the same housing. The control valve is located in the pump unit and meters the operating fluid at pump inlet. The rockshaft position can be automatically controlled by compression or tension on the upper implement attaching link, by a cam on rockshaft ram arm or by pressure in the ram cylinder. The system can be used to control rockshaft height (implement depth) or to transfer implement weight to rear tires for additional traction. The system differs from earlier Massey-Ferguson systems in that the control valve is spring-loaded toward the intake, or raising, side; and by the use of a dashpot to control implement response, rather than valve position.

A gear type auxiliary pump is optionally added to supply pressure for control of the "Multipower" and IPTO clutches; and/or pressure and flow for remote hydraulic cylinder applications.

The transmission lubricant is the operating fluid for the hydraulic system. Massey-Ferguson M-1129 Fluid is recommended. System capacity is 8 gallons.

TROUBLE SHOOTING

All Models

239. **SYSTEM CHECKS.** Before attaching an implement to tractor, start the engine, move response control lever to "Fast" position and position control lever to "Transport" position. With engine running at slow idle speed, check to make sure that rockshaft moves through full range of travel as draft control lever is moved to "Down" and "Up" positions. Using the draft control lever, stop and hold

the movement with lower links in an approximately horizontal position, lever should be centered between the sector marks (M—Fig. 269) on quadrant.

Attach an overhanging implement such as a plow to the links. With draft control lever in "Up" position and engine running at slow idle speed, raise and lower the implement a little at a time, using the position control lever. Implement should move in response to the lever and hold steady after completion of movement, through full range of rockshaft travel. Move lever to "Transport" position and scribe a line across lift arm hub and lift cover. Move position control lever past the transport stop to "Constant Pumping" position. The scribed lines should be separated 1/16-1/8 inch and relief valve should sound.

Hold the implement clear of the ground using the draft control lever (Position lever in "Transport"). If draft control spring is properly adjusted, system should respond to the application of pressure or lifting force to rear of implement, lowering when pressure is applied and raising when rear of implement is lifted.

Check system pressure and flow as outlined in paragraph 240. If system fails to perform as indicated, adjust as outlined in the appropriate following paragraphs and/or overhaul the system.

240. **PRESSURE AND FLOW.** Two different main hydraulic pumps have been used. On early tractors, relief valve setting was 2350 psi and rated delivery was 4.5 gpm @ 2000 engine rpm & 1500 psi. On late models, relief pressure has been increased to 2550 psi and pump delivery to 4.8 gpm. On models with pressure control, the system relief valve has been eliminated and maximum system pressure is determined by pressure control mechanism adjustment as outlined in paragraph 243.

The gear-type auxiliary pump may be any of several types, depending on date of manufacture and auxiliary uses. Only those with auxiliary hydraulic system affect the hydraulic lift system adjustments and tests. Refer also to paragraph 186 for pressure tests on "Multipower" and IPTO hydraulic circuits. On early models with auxiliary hydraulic system and "Multipower", the gear-type pump supplied a priority flow of 2 gpm to the "Multipower" control valve and 7 gpm @ 2000 engine rpm to the auxiliary hydraulic system. Late models with auxiliary hydraulic system plus "Multipower" or Independent Power Take-Off use dual hydraulic pump, with the small set of gears supplying 4 gpm to the "Multipower" and/or IPTO control circuits and the large set of gears supplying 8 gpm to the auxiliary hydraulic system. A

Fig. 269—Hydraulic lift control quadrant typical of that used on all models.

D. Draft control lever
M. Sector marks
P. Position control lever
R. Draft response lever

Fig. 270—Lever support bracket and internal linkage showing points of adjustment.

A. Adjusting screw
L. Locknut
N. Locknut
P. Port plug
S. Adjusting screw
T. Adjusting tube

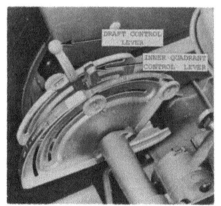

Fig. 271—View of hydraulic control quadrant used on models equipped with Pressure Control.

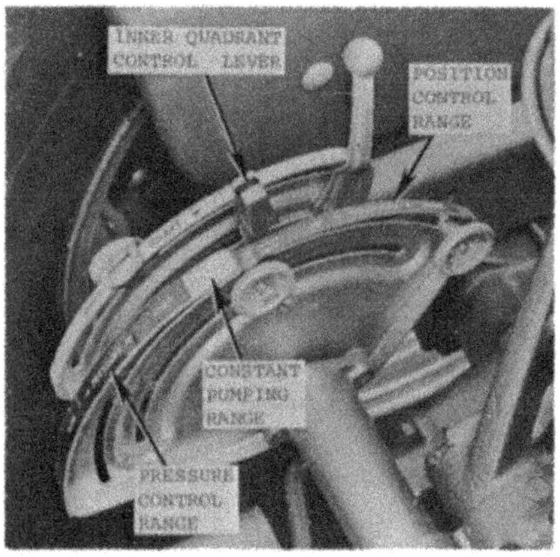

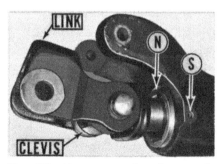

Fig. 273—Master control spring can be checked by disconnecting the link from lift cover, then pushing and pulling on clevis. To make an adjustment or remove the spring, loosen set screw (S) and turn nut (N).

Fig. 272—Inner quadrant control lever operates the position control range and pressure control range as shown.

combining valve is optionally available to combine the flow from both hydraulic system pumps for auxiliary or remote hydraulic use.

On models with Pressure Control, main pump pressure can be tested by installing a suitable pressure gage in pipe plug opening (P—Fig. 270) in hydraulic lift cover. With gage installed and engine running at slow idle speed, move draft control lever (Fig. 271) to top of quadrant as shown in Fig. 272. Move inner quadrant control lever fully forward to low end of pressure control range. Gage pressure should read approximately 20 psi.

Move inner quadrant control lever up pressure control range a little at a time, while noting gage pressure. Reading should increase at an even rate as lever is moved, until lever reaches "Constant Pumping" range. Maximum reading should be 2500-2650 psi for Models MF135, MF150 Std. & MF150 High Clearance Gasoline. Maximum pressure should be 2950-3100 psi for MF150 High Clearance Diesel and all MF165. If adjustment is not as indicated, adjust as in paragraph 243 or overhaul as in paragraph 253.

Auxiliary pump pressure and flow can be tested by installing suitable test equipment in breakaway couplings. On models equipped with the combining valve, pressure and flow of both systems can be tested by connecting a flow meter to breakaway couplings. Tests should be conducted at rated engine speed (2000 rpm) with transmission and hydraulic fluid at operating temperature, proceed as follows:

Move inner quadrant control lever to "Down" in Position Control Range or close the combining valve. Move auxiliary valve lever to "Raise" position and check auxiliary pump flow

which should be approximately 7 gpm for early models or 8 gpm for models with late pump. Move inner quadrant control lever to "Constant Pumping" position and/or open the combining valve and check the combined flow.

Close the restrictor valve on flow meter while observing pressure and flow. Relief pressure of the main hydraulic pump will be indicated when flow drops. To check the relief pressure of auxiliary pump, isolate the main pump by closing the combining valve.

ADJUSTMENTS
All Models

241. **CONTROL SPRING ADJUSTMENT.** To check the master control spring, disconnect the master control spring link as shown in Fig. 273 and check for end play by pushing and pulling on the clevis. If end play is present, loosen the Allen head set screw (S) in side of housing and unscrew the retainer nut (N), using the special spanner wrench FT-358. Withdraw the master control spring assembly as shown in Fig. 274. Remove the groove pin (P) and turn the clevis on control spring plunger until spring is snug but can still be rotated by hand pressure. Align one of the slots and insert groove pin.

Reinstall control spring assembly and turn retainer nut into housing until all end play is eliminated. NOTE: End play will exist if nut (N—Fig. 273) is either too tight or too loose. Tighten Allen head set screw (S) to a torque of 40-65 inch-pounds and complete the assembly by reversing the disassembly procedure.

242. **VALVE SYNCHRONIZATION.** The control valve, located in the pump body, must be synchronized with the control linkage located on the top cover, whenever either unit is removed, or as a check whenever trouble exists. To synchronize the linkage, drain the system down until the side cover containing the response control can be removed, then remove the cover.

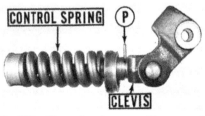

Fig. 274—Control spring assembly removed from cover. Remove pin (P) to turn clevis for spring adjustment.

Start and idle the engine, move position control lever to "Transport" position and center draft control lever between the two sector marks on quadrant. Reaching through side opening, move valve control lever (L—Fig. 275) forward and allow lift arms to lower, then release the lever and allow it to rest against adjusting screw (S). Thread adjusting screw (S) into control arm (A) until lift arms start to raise, then back the screw out until lift arms are stationary.

Check for proper response to draft lever setting by moving the lever to "Up", "Down" and neutral position as outlined in paragraph 239; then check (and readjust if necessary) the transport stop setting.

243. PRESSURE ADJUSTMENT. Main pump relief pressure is controlled by the same valve which regulates the weight transfer (Pressure Control) hitch. Adjustment must be made after valves are synchronized as outlined in paragraph 242 or if pressures were incorrect when tested as in paragraph 240.

To make the adjustment, remove plug (P—Fig. 270) from either side of lift cover and install a 3000 psi capacity pressure gage. Move draft control lever to transport position and inner quadrant control lever to "Constant Pumping" position. Start and idle the engine, then turn pressure

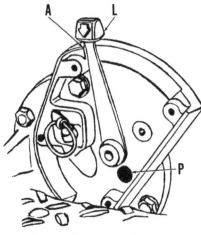

Fig. 276—To adjust the draft response, remove control cover as shown. Move lever (L) until distance (A) from rear stop measures 3/16-inch and remove plug from port (P) to make the adjustment.

control screw (P—Fig. 275) in or out until lever (L) starts to hunt or waver between intake and exhaust. At this point, back screw (P) out until gage pressure rises and holds steady. The setting is correct when diaphragm plunger of control (C) is in contact with screw (P) and the highest even pressure is obtained.

Gage pressure should be 3100 psi for MF165 or MF150 High Clearance Diesel; or 2650 psi for other models. If it is not, turn pressure adjusting tube (T—Fig. 270) either way as required until the correct reading is obtained.

244. RESPONSE ADJUSTMENT. To adjust the response control, remove response control cover and plug as shown in Fig. 276 and move lever (L) toward the slow position until distance (A) from rear stop measures 3/16-inch. Reaching through plug port (P), loosen the dashpot plunger adjusting screw (D—Fig. 275) slightly to release plunger rod, then retighten screw to a torque of 2-3 ft.-lbs.

NOTE: Dashpot plunger adjusting screw is an Allen head screw in some models, a hex head cap screw on others. Check to make sure which type of screw is used before attempting to make the adjustment.

245. TRANSPORT STOP ADJUSTMENT. On Models without pressure control, the transport stop (T—Fig. 277) should be adjusted to provide a full range of operation for the lift system rockshaft, allowing the control valve to return to neutral in the transport position, yet still provide

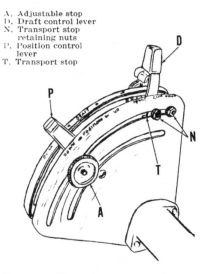

A. Adjustable stop
D. Draft control lever
N. Transport stop retaining nuts
P. Position control lever
T. Transport stop

Fig. 277—View of control quadrant used on models without pressure control, showing transport stop.

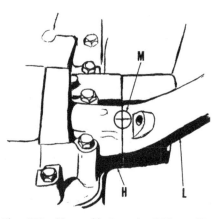

Fig. 278—Place chisel marks (M) on lift arm (L) and cover (H) with rockshaft in transport position. Refer to text.

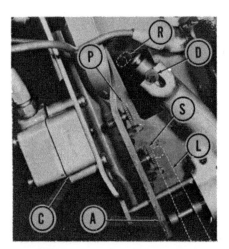

Fig. 275—Internal hydraulic control linkage showing points of final adjustment. Dotted lines indicate approximate positions of draft response cam (R) and valve control lever (L).

A. Control arm
C. Control diaphragm
D. Dashpot adjusting screw
L. Valve control
    lever
P. Pressure control screw
R. Respone cam
S. Adjusting screw

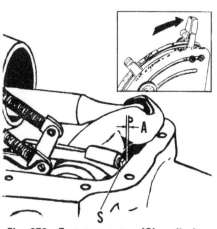

Fig. 279—Turn stop screw (S) until clearance (A) measures 0.160-0.180. Draft control lever must be in "Up" position as shown in inset.

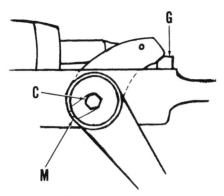

Fig. 280—Use gage block (G) or chisel marks (M) to position rockshaft in transport position. Refer to text.

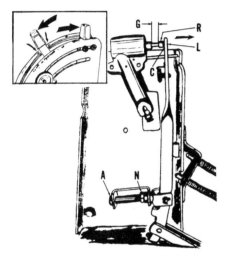

Fig. 281—Adjusting draft control linkage. Inset shows proper position of control levers on models without pressure control.

A. Adjusting screw
C. Clearance (0.000-0.002)
G. Gap
L. Control arm
N. Locknut
R. Dashpot plunger

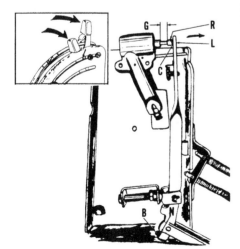

Fig. 282—Adjusting position control linkage. Inset shows proper setting of control levers on models without pressure control.

B. Adjusting screw
C. Clearance (0.000-0.002)
G. Gap
L. Control arm
N. Locknut
R. Dashpot plunger

for a constant pumping position for auxiliary hydraulic use.

To adjust the transport stop, start the engine and run at slow idle speed. Move the position control lever (P) past transport stop (T) to "Constant Pumping" position. Engine should labor slightly and hydraulic pressure relief valve should sound. With position control lever in constant pumping position, scribe a line across the junction of one lift arm (L—Fig. 278) and housing (H) as shown at (M). Move the lever (P—Fig. 277) back down the quadrant, then back up to a position resting against the transport stop (T); the two marks (H—Fig. 278) should now be separated by a distance of 1/16-⅛ inch, if they are not, loosen the two nuts (N—Fig. 277) and slide the stop (T) on quadrant until the correct adjustment is obtained. With transport stop correctly positioned and lift arms in fully raised ("Transport") position, place chisel marks (M—Fig. 278) on housing (H) and lift arm (L) in a convenient place for future reference.

246. INTERNAL LINKAGE ADJUSTMENT. To adjust the internal lift linkage, first check to make sure lift arm transport position is marked as outlined in paragraph 245, then remove hydraulic lift cover as in paragraph 247. Check and/or adjust the control spring as outlined in paragraph 241.

Invert the lift cover in a fixture or on a bench, blocking up the cover so that rockshaft can be moved to the normal raised "Transport" position. Move draft control lever to full "Up" position and, using Massey-Ferguson special gage MFN-124 or an 11/64-inch drill bit as a gage, adjust the clearance (A—Fig. 279) to 0.160-0.180 by turning stop screw (S) as required.

Position the rockshaft in transport position using the special Massey-Ferguson tool MFN-970, as shown at (G—Fig. 280). NOTE: Two different lift covers are used; before positioning the gage, measure front flange thickness of cover. If front flange is ⅞-inch thick, side of gage marked ".470" must face ram arm. If front flange is 1⅛-inches thick, side marked ".636" must face ram arm. If the special gage is not available, align the previously scribed transport position alignment marks (M—Fig. 278), then tighten cap screw (C—Fig. 280) (or screws) in one end of rockshaft until shaft will maintain a fixed position with marks (M) aligned.

To adjust the draft control linkage, refer to Fig. 281 and proceed as follows: Move position control lever to "Constant Pumping" position and center the draft control lever between sector marks on quadrant as shown in inset. Block the draft response plunger (R) in fully extended neutral position by inserting the Massey-Ferguson special wedge (MFN-163) or other suitable wedge in gap (G) between plunger (R) and plunger guide. Apply a light (approximately 3 lb.) pull to end of vertical control lever (L) in direction indicated by arrow, loosen locknut (N) and turn adjusting bolt (A) in or out as required until clearance (C) between lever (L) and plunger (R) is 0.000-0.002. Lock the adjustment by tightening locknut (N).

To adjust the position control linkage, refer to Fig. 282. Move draft control lever to "Up" position and position control lever into contact with transport stop as shown in inset and block the draft response plunger (R) in extended position as indicated for draft linkage adjustment. Apply a 3 lb. pull to end of vertical lever (L) in direction of arrow and turn adjusting screw (B) in or out as required until clearance (C) between plunger (R) and lever (L) is 0.000-0.002. Tighten locknut (N) to secure the adjustment. Refer to Fig. 270 for view of linkage on models with pressure control.

Synchronize the linkage with control valve after cover is installed, as outlined in paragraph 242.

LIFT COVER
All Models

247. REMOVE AND REINSTALL. To remove the hydraulic lift cover, first remove seat and seat frame and disconnect upper lift links from lift arms. On models so equipped, disconnect control beam (6—Fig. 284) from clevis (16). On all models, drain hydraulic system fluid down until response side cover can be removed and remove the cover. Reaching through side opening, spread valve control lever arms (L—Fig. 283) and remove roller (R).

Remove transfer plate (3—Fig. 284) or auxiliary valve. On early models without pressure control, withdraw standpipe (2). On early models with pressure control and all late models, the standpipe contains a welded tee-block and cannot be withdrawn from above.

Remove the attaching cap screws, then using suitable hoisting equipment, carefully remove the cover. On models with pressure control, if upper end of pressure tube (2—Fig. 289) is square cut, grind a 15° chamfer to lead into O-ring in standpipe tee block.

Keep cover level when installing. On models with welded tee block in standpipe, install standpipe in lift cover from bottom and push pipe up as far as possible. Make sure linkage support bracket on cover is positioned between arms (L—Fig. 283) of valve lever. Tighten the cover retaining cap screws to a torque of 50-55 ft.-lbs., reinstall valve lever roller (R) and complete the installation by reversing the removal procedure. Synchronize valve linkage after cover is installed, as outlined in paragraph 242.

248. **OVERHAUL.** Refer to Figs. 284, 285, 286 and 287 for exploded views of lift cover and linkage. Lift cover also contains draft response dashpot (and pressure control valve if so equipped). Dashpot and control valve may be removed and overhauled at this time as outlined in paragraphs 249 and 250.

Lift arms, ram arm and rockshaft all have master splines for correct assembly. Rockshaft bushings are a slip fit in top cover bores.

To remove the control quadrant and associated parts, first remove the countersunk set screw (R—Fig. 284) then withdraw the unit as an assembly. With quadrant out, control cams can be withdrawn after loosening the retaining set screw and withdrawing the pivot shaft.

NOTE: Guide rods for control cam return springs (34—Fig. 286) are drilled near loose end to assist in dis-

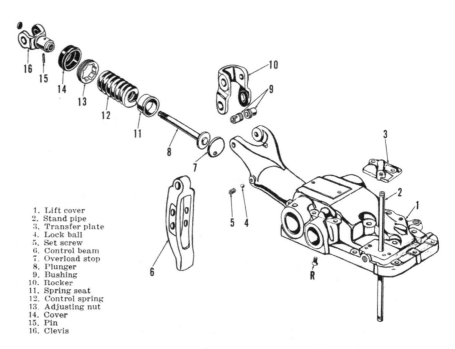

1. Lift cover
2. Stand pipe
3. Transfer plate
4. Lock ball
5. Set screw
6. Control beam
7. Overload stop
8. Plunger
9. Bushing
10. Rocker
11. Spring seat
12. Control spring
13. Adjusting nut
14. Cover
15. Pin
16. Clevis

Fig. 284 — Hydraulic lift cover showing master control spring and associated parts. Countersunk screw (R) retains control quadant.

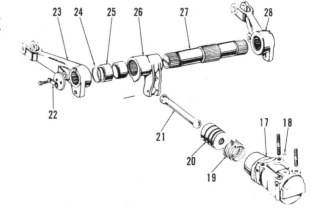

Fig. 285—Exploded view of ram cylinder, rockshaft and associated parts.

17. Cylinder
18. "O" ring
19. Piston rings
20. Ram piston
21. Connecting rod
22. Lock clip
23. Lift arm
24. "O" ring
25. Bushing
26. Ram arm
27. Rockshaft
28. Lift arm

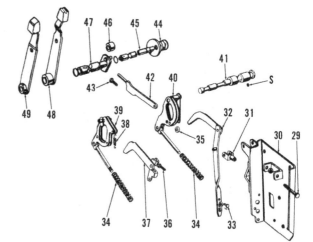

Fig. 286—Hydraulic lift internal control linkage.

29. Adjusting bolt
30. Support bracket
31. Pivot yoke
32. Control arm
33. Adjusting screw
34. Spring
35. Cam roller
36. Adjusting screw
37. Position control arm
38. Return spring
39. Position control cam
40. Draft control cam
41. Pivot shaft
42. Draft control rod
43. Stop screw
44. Cam roller
45. Draft control shaft
46. Cam roller
47. Position control shaft
48. Position control lever
49. Draft control lever
S. Set screw

Fig. 283—Control valve lever (L) and roller (R) assembled to pump unit.

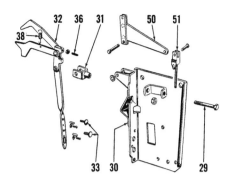

Fig. 287—Hydraulic lift internal control linkage used on models with Pressure Control.

29. Adjusting bolt
30. Support bracket
31. Pivot yoke
32. Control arm
33. Adjusting screws
36. Adjusting screw
38. Return spring
50. Pressure control lever
51. Adjusting rod

assembly and assembly. Compress the springs and insert cotter pins through drilled holes to retain the springs.

Remove master control spring and associated parts as outlined in paragraph 241. Refer to paragraph 249 for overhaul of response control dashpot and to paragraph 250 for service on pressure control valve. Assemble the cover by reversing the disassembly procedure and adjust as outlined in paragraphs 241 through 246.

249. DRAFT RESPONSE DASHPOT. The draft response dashpot assembly (Fig. 288) attaches to lever bracket (30—Fig. 286 or 287), and can be removed after removing lift cover as outlined in paragraph 247.

To disassemble the dashpot, loosen adjusting screw (18 or 18A—Fig. 288) and remove adjustment rod (20) and spring (19). Invert the housing and remove plunger (17), needle (15) and spring (14). Ball (16) was used on some models only. Depress piston rod and guide (8) and remove snap ring (7), then remove spring (9), piston

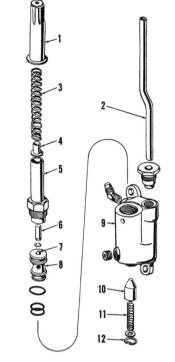

Fig. 289—Exploded view of pressure control valve and associated parts.

1. Adjusting tube
2. Pressure tube
3. Pressure spring
4. Spring guide
5. Pilot
6. Plunger
7. Piston
8. Piston sleeve
9. Body
10. Valve
11. Spring
12. Snap ring

(10) and spring (11). Expansion plugs or guide (12) can be removed from body if renewal is indicated.

Assemble by reversing the disassembly procedure, using Fig. 288 as a

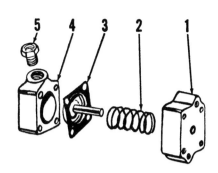

Fig. 290—Exploded view of pressure control diaphragm.

1. Cover
2. Spring
3. Diaphragm
4. Body

guide. Adjust as outlined in paragraph 244 after tractor is assembled.

250. PRESSURE CONTROL VALVE. The pressure control valve shown exploded in Fig. 289 serves the dual purpose of providing pressure relief for the main hydraulic pump during normal operation; and providing weight transfer for added traction when using some types of pulled or mounted implements.

Pressure line (2) is connected to welded tee-block in standpipe, and ram cylinder pressure acts against servo piston (7) which is held closed by the variable rate spring (3). The spring is compressed to the maximum when inner quadrant control lever is in "Position Control" or "Constant Pumping" position, and is released at a uniform rate as lever is moved to low end of "Pressure Control" sector of quadrant. When pressure valve is correctly adjusted and operating properly, pressure in the ram cylinder circuit should be 20-40 psi with inner quadrant control lever at extreme "Low" end of quadrant and should increase at a steady rate as lever is moved, until the specified relief pressure is obtained as lever moves into "Constant Pumping" sector of quadrant.

The servo piston (7) and matched sleeve (8) connect the control diaphragm assembly (Fig. 290) to control passage of pressure valve. When servo piston is against stop pin in valve body (9—Fig. 289) the diaphragm passage is open to reservoir. As pressure builds up in ram cylinder circuit and piston (7) moves upward against spring pressure, the diaphragm exhaust passage is closed. When the pressure limit is reached, fluid passes servo piston to diaphragm passage, extending the diaphragm

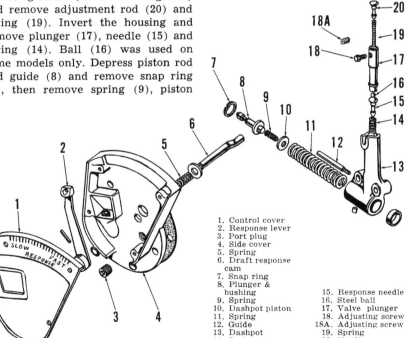

1. Control cover
2. Response lever
3. Port plug
4. Side cover
5. Spring
6. Draft response cam
7. Snap ring
8. Plunger & bushing
9. Spring
10. Dashpot piston
11. Spring
12. Guide
13. Dashpot
14. Spring
15. Response needle
16. Steel ball
17. Valve plunger
18. Adjusting screw
18A. Adjusting screw
19. Spring
20. Needle

Fig. 288—Draft response dashpot, control cover and associated parts.

Fig. 291—Side of rear axle center housing showing hydraulic pump positioning dowel pin (P).

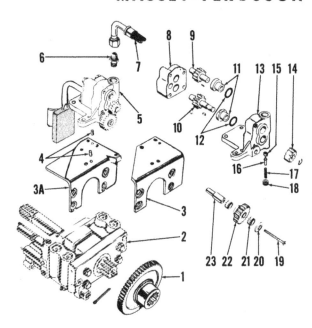

Fig. 293—Exploded view of "Multipower" pump and associated parts.

1. Drive gear
2. Main hydraulic pump
3. Mounting bracket (early)
3A. Mounting bracket (late)
4. Dowel
5. Multipower pump
6. Adapter
7. Pressure hose
8. Pump cover
9. Follow gear
10. Driven gear
11. Bushings
12. Seal ring
13. Pump body
14. Drive gear
15. Valve seat
16. Relief valve
17. Spring
18. Adjusting screw
19. Retaining screw
20. Retaining washer
21. Needle bearing
22. Idler gear
23. Idler shaft

plunger to move the main hydraulic pump control arm to neutral.

Relief valve (10) is pre-set at a slightly higher pressure than diaphragm return spring (2—Fig. 290). The valve provides a safety relief passage if diaphragm plunger malfunctions or system is improperly adjusted.

To disassemble the pressure control valve, refer to Fig. 289 and proceed as follows: Withdraw adjusting tube (1), spring (3) and guide (4), then unscrew and remove pilot (5). Using needle nose pliers, carefully withdraw servo piston (7) and remove piston sleeve (8) using a hooked wire. Relief valve (10) and spring (11) can be removed after removing snap ring (12). Servo piston (7) and sleeve (8) are available only as a matched set. All other parts are available indi-

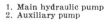
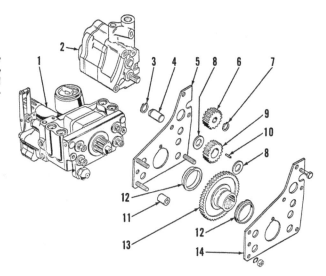

Fig. 294—Exploded view of auxiliary pump drive gear train and associated parts used on late models

1. Main hydraulic pump
2. Auxiliary pump
3. Snap ring
4. Idler shaft
5. Mounting bracket
6. Pump gear
7. Snap ring
8. Thrust washer
9. Idler gear
10. Needle roller
11. Spacer
12. Bushing
13. Drive gear
14. Bracket

vidually. Examine all parts carefully and renew any which are worn, scored or damaged. Assemble the pressure control valve be reversing the disassembly procedure. Adjust as outlined in paragraph 243 after lift cover is reinstalled.

MAIN HYDRAULIC PUMP
All Models

251. REMOVE AND REINSTALL. To remove the hydraulic system pump on models not equipped with auxiliary pump, first drain system and remove lift cover as outlined in paragraph 247. Collapse and remove the drive shaft coupler and remove pto output shaft. Remove the dowel pin (P—Fig. 291) from each side of rear axle center housing, then lift hydraulic pump out through top opening.

Fig. 292—Reposition the pump package and remove elbow (1) when removing the unit.

1. Elbow
2. Auxiliary pump
3. Drive gear

Fig. 295—Hydraulic pump with valve lever removed. Refer to Fig. 297 for parts identification.

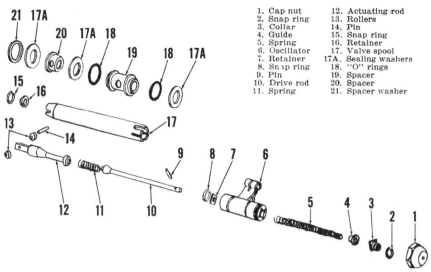

1. Cap nut	12. Actuating rod
2. Snap ring	13. Rollers
3. Collar	14. Pin
4. Guide	15. Snap ring
5. Spring	16. Retainer
6. Oscillator	17. Valve spool
7. Retainer	17A. Sealing washers
8. Snap ring	18. "O" rings
9. Pin	19. Spacer
10. Drive rod	20. Spacer
11. Spring	21. Spacer washer

Fig. 296—Exploded view of hydraulic pump control valve and camshaft driven oscillator which prevents valve from sticking. Nut (1) is shown in proper relation to pump housings, in Fig. 297.

Fig. 298 — Disassembled view of pump valve chamber showing poppet type intake and exhaust valves.

On models equipped with auxiliary pump, drain the system and remove hydraulic lift cover as outlined in paragraph 247 and transmission top cover as in paragraph 175. Remove step plates and disconnect brake rods & multipower pressure line (if so equipped). Support transmission and rear axle center housing separately, remove flange bolts and separate tractor at rear of transmission housing. Disconnect and remove auxiliary pump hydraulic lines leading to side cover and IPTO clutch valve on models so equipped. Remove pump

mounting dowel (P—Fig. 291) from each side of rear axle center housing and slide pump package forward through front opening of rear axle center housing.

NOTE: On some models with auxiliary hydraulic system, it is necessary to reposition pump and remove elbow (1—Fig. 292) before pump package can be withdrawn.

A number of changes have been made in main hydraulic pump, auxiliary pump and auxiliary pump mounting brackets. Fig. 293. On early models using bracket (3), backlash adjustment of auxiliary pump drive gears was made by loosening the four

stud nuts securing bracket to hydraulic pump (2) and shifting bracket on pump. On models using bracket (3A), shims are interposed between auxiliary pump (5) and mounting bracket. The pump mounting shims can be used for backlash adjustment on earlier models. Shims are available in thicknesses of 0.002, 0.005 and 0.010, and recommended backlash is 0.002-0.005.

On late models, the gear train mounting plates serve as pump support. Refer to Fig. 294. Gear train must be disassembled to remove pump. Idler gear (9) contains 22 loose needle bearings (10). Gear backlash is not adjustable; renew parts concerned if backlash exceeds 0.015 be-

1. Cap nut	
22. Coupler	
22A. Gear & Coupler	
23. Bushing	
24. Front cover	
25. Valve chambers	
26. Cam block	
27. Oscillator drive	
28. Piston	
29. Piston rings	
30. Cam block	
31. Camshaft	
32. Needle bearing	37. Filter cover
33. Rear body	38. Gasket
34. Intake housing	39. Rear cover
35. Filter bowl	40. Push stud
36. Filter element	41. Actuating lever
	42. Roller
	43. Relief valve

Fig. 297—Exploded view of hydraulic system pump and associated parts.

Fig. 299—Use a suitable ring compressor as shown when installing side chamber.

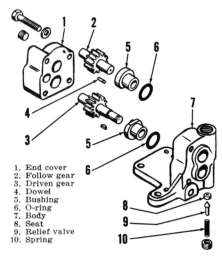

1. End cover
2. Follow gear
3. Driven gear
4. Dowel
5. Bushing
6. O-ring
7. Body
8. Seat
9. Relief valve
10. Spring

Fig. 300—Exploded view of Wooster 4 qpm pump used on early models with Multipower transmission.

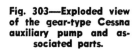

Fig. 303—Exploded view of the gear-type Cessna auxiliary pump and associated parts.

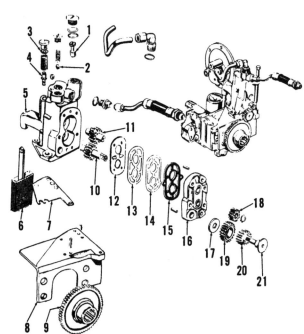

1. Flow divider spool
2. Priority relief valve
3. Adjusting shims
4. Auxiliary relief valve
5. Pump body
6. Intake screen
7. Mounting shim pack
8. Mounting bracket
9. Gear & coupler
10. Driven gear
11. Follow gear
12. Diaphragm
13. Gasket
14. Protector
15. Seal
16. Cover
17. Washer
18. Drive gear
19. Idler gear
20. Needle rollers
21. Idler stud

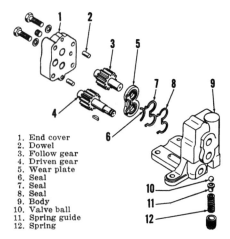

1. End cover
2. Dowel
3. Follow gear
4. Driven gear
5. Wear plate
6. Seal
7. Seal
8. Seal
9. Body
10. Valve ball
11. Spring guide
12. Spring

Fig. 301—Exploded view of 4 gpm service pump available through parts stock.

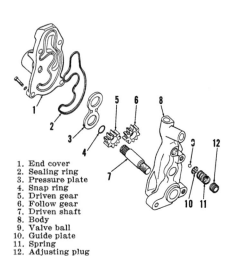

1. End cover
2. Sealing ring
3. Pressure plate
4. Snap ring
5. Driven gear
6. Follow gear
7. Driven shaft
8. Body
9. Valve ball
10. Guide plate
11. Spring
12. Adjusting plug

Fig. 302—Exploded view of Plessey 4 gpm pump used on late models so equipped.

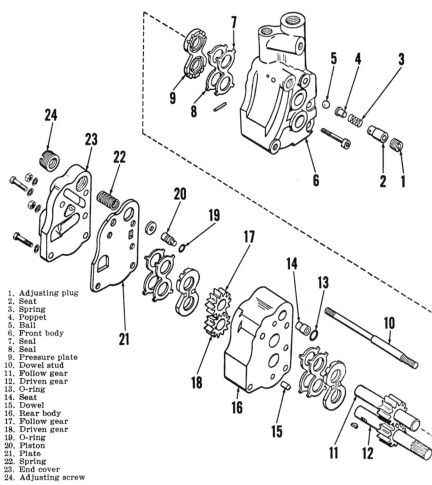

1. Adjusting plug
2. Seat
3. Spring
4. Poppet
5. Ball
6. Front body
7. Seal
8. Seal
9. Pressure plate
10. Dowel stud
11. Follow gear
12. Driven gear
13. O-ring
14. Seat
15. Dowel
16. Rear body
17. Follow gear
18. Driven gear
19. O-ring
20. Piston
21. Plate
22. Spring
23. End cover
24. Adjusting screw

Fig. 304—Exploded view of Warner Motive Dual Section pump used on late models with auxiliary hydraulics.

tween any two gears. Tighten retaining bolts to a torque of 30-35 ft.-lbs.

252. **OVERHAUL.** To disassemble the removed hydraulic pump, pull straight out on actuating lever (41—Fig. 295) to free lever from retaining stud (40), then lift the lever off of actuating rod (12) and rollers (13). Remove coupling (22—Fig. 297) or drive gear (22A) on models not equipped with late auxiliary pump drive. Remove nut (1) and unbolt and remove pump cover (24). Remove actuating pin (14—Fig. 296) and rollers (13) and disconnect oscillator (6) from cam follower (27—Fig. 297). Remove control valve and oscillator assembly from the pump.

The control valve is serviced as a matched assembly which includes the spool (17) and three sealing washers (17A). To disassemble the control valve spool, first remove retaining ring (2) and pin (9) and remove oscillator (6) and associated parts; then remove snap ring (15), retainer (16), actuating rod (12), spring (11) and oscillator drive rod (10) from spool. To remove the sealing washers (17A), unbolt and remove rear cover (39—Fig. 297) and intake housing (34) from rear body (33) and push sealing washers (17A—Fig. 296), spacers (19 & 20), O-rings (18) and spacer washer (21) from rear body. Assemble by reversing the disassembly procedure, using Figs. 296 and 297 as a guide.

Withdraw side chambers (25—Fig. 297) from pistons (28). Side chambers contain poppet type inlet and exhaust valves as shown in Fig. 298. Valves and associated parts can be removed for cleaning or parts renewal after removing the retaining snap rings. Use a suitable ring compressor as shown in Fig. 299 when installing pistons in valve chambers. Tighten pump retaining stud nuts to a torque of 30-35 ft.-lbs.

AUXILIARY PUMP
All Models So Equipped

253. A number of different auxiliary pumps have been used. Refer to Figs. 300 through 304. Pumps shown in Fig. 300, 301 and 302 are used on tractors equipped only with "Multipower" transmission, Independent Power Take-Off, or a combination of the two. Pumps shown in Fig. 303 and 304 are used for auxiliary hydraulic applications as well as for "Multipower" or IPTO pressure.

WOOSTER 4 gpm pump: Refer to Fig. 300. End bearings (5) are pressure loaded and balanced to control

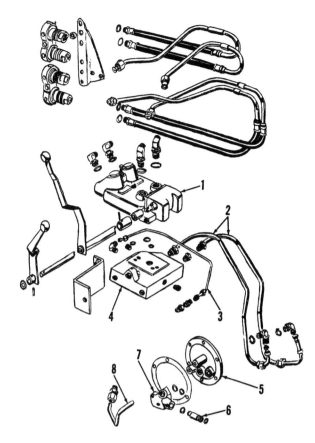

Fig. 305—Dual auxiliary valve, manifold plates and associated parts.

2. Valve lines
3. Bleed line
4. Manifold
5. Side cover
6. Pump inlet line
7. Manifold
8. Pump auxiliary line

gear end clearance. Note location of bearings and gears and return them to same relative position when reassembling. Bearings, gears and pressure relief valve components are available individually. If body (7) or end cover (1) are damaged, renew the pump.

WARNER MOTIVE 4 gpm pump: Refer to Fig. 301. Wear plate (5) is pressure loaded and sealed to control gear end clearance. Body (9) is available from parts stock but end cover (1) is not. All other parts are available individually.

PLESSEY 4 gpm pump: Refer to Fig. 302. Pressure plate (3) is pressure loaded to control gear end clearance. Pressure plate and the two pumping gears are available as a matched kit. End cover (1) is available separately but pump body (8) is not. Bearings, shaft, seal and pressure relief valve components are available individually.

CESSNA Dual Delivery 9 gpm pump: Refer to Fig. 303. The sleeve for priority valve (1) and seat for poppet valve (4) are factory installed and removal is not recommended during pump service. With the exception of priority valve (1) and pump

body assembly (5), all parts shown are available individually. When assembling the pump, use a blunt pointed tool and install V-seal (15) in grooves in cover (16) with open side down. Make sure V-seal is completely seated. Install protector gasket (14), backup gasket (13) and diaphragm (12), with bronze side next to gears. Tighten the cap screws retaining cover (16) evenly to a torque of 7-10 ft.-lbs.

WARNER MOTIVE Dual Section 8/4 gpm pump: Refer to Fig. 304. To disassemble the pump, remove cap screws and stud nut securing end cover (23); cover will be forced off by pressure of spring (22). Do not remove high pressure adjusting screw (24); screw is staked in place and setting should not be changed. Pry rear body (16) from front body (6) using a suitable tool in the notches provided. If pump housings show evidence of wear or scoring in gear pocket area, renew the pump. If housings are serviceable, gears, relief valves and seals can be renewed. When assembling the pump, omit high pressure relief valve spring (22) and its washer, and use the body cap screws to draw bodies together over dowel pins. Remove screws and cover and install relief valve spring. Tighten to a torque of 18-20 ft.-lbs. When

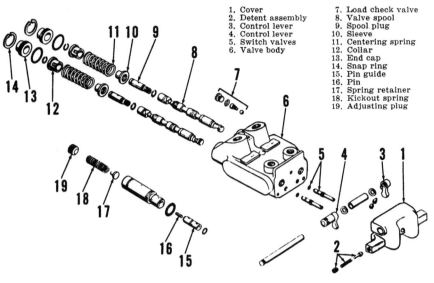

1. Cover	7. Load check valve
2. Detent assembly	8. Valve spool
3. Control lever	9. Spool plug
4. Control lever	10. Sleeve
5. Switch valves	11. Centering spring
6. Valve body	12. Collar
	13. End cap
	14. Snap ring
	15. Pin guide
	16. Pin
	17. Spring retainer
	18. Kickout spring
	19. Adjusting plug

Fig. 306—Exploded view of auxiliary dual control valve. A single control valve of the same type may be used.

installing low pressure relief valve, (parts 2 through 5), tighten adjusting plug (1) until it bottoms, back out four full turns and stake in place, to establish relief pressure within the recommended range of 650-800 psi.

AUXILIARY VALVE
All Models So Equipped

254. Refer to Fig. 305 and Fig. 306 for exploded views. The single remote valve is similar to one half of the dual valve shown.

COMBINING VALVE
All Models So Equipped

255. Refer to Fig. 307 for an exploded view of valve. The unit combines the flow of the main and auxiliary hydraulic pumps providing a fluid flow of approximately 12 gpm for use of hydraulic equipment.

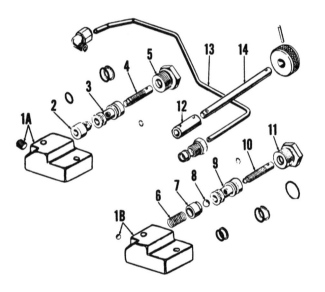

Fig. 307—Exploded view of combining valve. Early type unit is shown at top-left.

1A. Body (early)	
1B. Body (late)	
2. Poppet	
3. Sleeve	
4. Stem	
5. Gland nut	
6. Spring	
7. Retainer	
8. Ball	
9. Sleeve	
10. Stem	
11. Gland nut	
12. Coupling	
13. Supply tube	
14. Rod	

NOTES

NOTES